Reading the Earth

Reading the Earth
New Directions in the Study of Literature and Environment

EDITED BY
*Michael P. Branch, Rochelle Johnson,
Daniel Patterson, and Scott Slovic*

University of Idaho Press
1998

Copyright © 1998 by the University of Idaho Press
Published by the University of Idaho Press,
Moscow, Idaho 83844–1107
Printed in Canada
All Rights Reserved

No part of this publication may be reproduced, stored in a retrieval system, or transmitted in any form or by any means, electronic, mechanical, photocopying, recording, or otherwise, except for purposes of scholarly review, without the prior permission of the copyright owner.

02 01 00 99 98   5 4 3 2 1

Library of Congress Cataloging-in-Publication Data

Reading the earth : new directions in the study of literature and environment / edited by Michael P. Branch . . . [et al.].
    p.   cm.
  Essays based on lectures and papers delivered at the first conference of the Association for the Study of Literature and Environment.
  Includes bibliographical references and index.
  ISBN 0-89301-213-0 (alk. paper)
  1. American literature—History and criticism. 2. Environmental literature—History and criticism. 3. English literature—History and criticism. 4. Environmental protection in literature. 5. Environmental policy in literature. 6. Ecology in literature. 7. Nature in literature.   I. Branch, Michael P.
PS169.E25R43   1998
810.9'355—dc21   97-39439
                                                                        CIP

Contents

Acknowledgments ix
Introduction xi

PART I
THEORETICAL PERSPECTIVES ON CULTURE
AND ENVIRONMENT

Ego or Eco Criticism? Looking for Common Ground 3
    WILLIAM HOWARTH

Toward an Ecology of Justice:
Transformative Ecological Theory and Practice 9
    JONI ADAMSON CLARKE

Talking about Trees in Stumptown:
Pedagogical Problems in Teaching EcoComp 19
    MICHAEL MCDOWELL

Dropping the Subject: Reflections on the Motives
for an Ecological Criticism 29
    ERIC TODD SMITH

Bodega Head: An Excursion in Nuclear Shamanism 41
    JOHN P. O'GRADY

PART II
GENRE, GENDER, AND THE BODY OF NATURE

"Whole Shoals of Men": Representations of Women Anglers in
Seventeenth-Century British Poetry 55
    ANNE E. MCILHANEY

Dorothy Wordsworth, Ecology, and the Picturesque 67
ROBERT MELLIN

Mary Austin's Nature: Refiguring Tradition through the Voices of Identity 79
ANNA CAREW-MILLER

Misogyny in the American Eden: Abbey, Cather, and Maclean 97
J. GERARD DOLLAR

The Body as Bioregion 107
DEBORAH SLICER

PART III
READINGS OF NINETEENTH-CENTURY
ENVIRONMENTAL LITERATURE

The Ornithological Autobiography of John James Audubon 119
CHRIS BEYERS

"A beautiful and thrilling specimen": George Catlin, the Death of Wilderness, and the Birth of the National Subject 129
DAVID MAZEL

Nathaniel Hawthorne Had a Farm: Artists, Laborers, and Landscapes in The Blithedale Romance 145
KELLY M. FLYNN

Agrarian Environmental Models in Ralph Waldo Emerson's "Farming" 155
STEPHANIE SARVER

Exploring the Linguistic Wilderness of The Maine Woods 165
ANN E. LUNDBERG

"I only seek to put you in rapport": Message and Method in Walt Whitman's Specimen Days 179
DANIEL J. PHILIPPON

PART IV
READINGS OF TWENTIETH-CENTURY
ENVIRONMENTAL LITERATURE

Beyond the Excursion: Initiatory Themes in Annie Dillard and Terry Tempest Williams 197
JOHN TALLMADGE

Aimé Césaire's A Tempest *and Peter Greenaway's* Prospero's Books *as Ecological Readings and Rewritings of Shakespeare's* The Tempest   209
   PAULA WILLOQUET-MARICONDI

Seeing, Believing, and Acting: Ethics and Self-Representation in Ecocriticism and Nature Writing   225
   H. LEWIS ULMAN

Don DeLillo's Postmodern Pastoral   235
   DANA PHILLIPS

"The world was the beginning of the world": Agency and Homology in A. R. Ammons's Garbage   247
   LEONARD M. SCIGAJ

Notes on Editors and Contributors   259
Index   263

Acknowledgments

It is our pleasure to acknowledge the direct and indirect support of many colleagues in the editing and production of this volume. We are grateful to Susan Beegel for her early interest in the project and for her efforts as a liaison between the Association for the Study of Literature and Environment (ASLE) and the University of Idaho Press. Throughout the publishing process we have received helpful support and advice from the staff at the University of Idaho Press.

Closer to home, we wish to acknowledge the University of Nevada, Reno, and California State University, San Bernardino, for their support of the editors of this volume. For his technical support of the project, our thanks to Brad Lucas. And a very special thanks to Susan M. Lucas of the Literature and Environment Program at the University of Nevada, Reno, for her outstanding professional assistance with the editing and preparation of this manuscript.

Of course, we are especially grateful to the twenty-one authors included in this volume for their fine work and for their exemplary patience in working through many rounds of drafts and queries. Finally, our thanks to the hundreds of ASLE members who enriched our own conceptions of ecocriticism by allowing us to read and consider their papers; as a token of our appreciation for their important work, the editors have donated all royalties generated by the sale of this volume directly to ASLE.

Introduction

The study of literature and environment, often referred to as ecocriticism, is rapidly building a momentum and legitimacy that attest to its exciting relevance and to its usefulness as a means of inquiry into the relationship between human culture and the nonhuman natural world. Over the past several years, an increasing number of scholars have solidified the theoretical basis for work in this area of literary studies, thereby nurturing a phenomenal increase in ecocritical activity. It is now clear that ecocriticism engages many important ideas that critics and teachers have long worked with in relative isolation from their colleagues. Recently, the emergence of a cohesive community of interested scholars has made possible a rich diversity of new directions in the study of literature and environment.

This collection of original essays exemplifies the range of approaches and the freshness of insight that distinguish the emerging field of ecocriticism. The book also suggests the ways in which creative, informed examination of the vital connections between literature and the physical environment can enrich the value of contemporary literary studies, both for academics and for general readers. While many of the basic tenets of and models for ecocriticism have been gathered into Cheryll Glotfelty and Harold Fromm's *The Ecocriticism Reader: Landmarks in Literary Ecology* (1996), the purpose of *Reading the Earth: New Directions in the Study of Literature and Environment* is to demonstrate several of the important new directions in this quickly expanding area of study. Ecocritic Lawrence Buell, as quoted in the 1996 article from the *Chronicle of*

Higher Education entitled "Inventing a New Field: The Study of Literature About the Environment," comments that "the worst thing that could happen would be for ecocriticism to become just another branch of literary criticism" (Winkler A15). The articles included in this volume show how ecocritics are testing the boundaries of academic discourse and exploring the cultural implications of literature—how, in other words, these scholars are working to avoid creating "just another branch of literary criticism."

Literary criticism since the 1950s has attempted ever broader and more diverse representation of authors and texts. The voices of women, African Americans, Native Americans, and other underrepresented groups have one after another been studied by scholars and brought into the classroom, and knowledge of these voices has become a prerequisite to cultural literacy in late-twentieth-century America. However, the "voice" longest neglected has been that of our physical environment, the voice of nature, which cannot speak through conventional means. The critical approaches represented in this volume respond to concerns about race, class, and gender, while also extending legitimacy and critical consideration to the long-unheard voices of the environment. Thus a new interdisciplinary field of inquiry, insight, and knowledge opens before the current generation of scholars, teachers, and writers amid a renaissance of environmental writing—a renaissance that is most active in the United States but which is fast becoming a formidable influence upon literature and literary studies in many other countries.

More than critical theory and ethical considerations account for the emergence of ecocriticism: the global environmental crisis demands a new approach to literary studies. David W. Orr offers one view of the potential danger of educational institutions not adequately responding to this crisis:

> By failing to include ecological perspectives in any number of subjects, students are taught that ecology is unimportant for history, politics, economics, society, and so forth. And through television they learn that the earth is theirs for the taking. The result is a generation of ecological yahoos without a clue why the color of the water in their rivers is related to their food supply, or why storms are becoming more severe as the planet warms. The same persons as adults will create businesses, vote, have families, and above all, consume. (85–86)

As a species humans have contributed little to the health of our planet, and the recent history of humans on earth has too often been one of taking, depleting, exhausting, trashing, and ignoring our environment. One important use of an ecologically sensitive literary criticism, then, lies in its potential to promote greater ecological literacy among all members of the human community. As one response to this global environmental crisis, ecocriticism suggests means by which we might read literary texts with a new appreciation for what they reveal about the complex of relationships that mediate interactions between humans and their environments. Environmentally informed literary scholarship offers a profound opportunity to read literature with a fresh sensitivity to the emergent voice of nature.

Implicit (and often explicit) in much of this new criticism is a call for cultural change. Ecocriticism is not just a means of analyzing nature in literature; it implies a move toward a more biocentric world-view, an extension of ethics, a broadening of humans' conception of global community to include nonhuman life forms and the physical environment. Just as feminist and African American literary criticism call for a change in culture—that is, they attempt to move the culture toward a broader world-view by exposing an earlier narrowness of view—so too does ecological literary criticism advocate for cultural change by examining how the narrowness of our culture's assumptions about the natural world has limited our ability to envision an ecologically sustainable human society. Barry Lopez has written that nature writing "will not only one day produce a major and lasting body of American literature, but . . . it might also provide the foundation for a reorganization of American political thought" (297). Ecocritical analysis is a clarifying voice in this process of cultural scrutiny and reshaping.

Beyond clearly environmental and political considerations, many practitioners of ecocriticism feel a need for a literary critical practice that recognizes, addresses, and embraces those aspects of literature that surpass and subsume both the political and the rational. An ecocritical orientation often leads beyond these tangible matters to a deeper concern with desire and wisdom—to a consideration of how a text resists, expresses, or inspires what biologist E. O. Wilson has called the "biophilic" desire of the human species; according to Wilson, "biophilia" may be defined as "the innate [human] tendency to focus on life and lifelike processes," the impulse of fascination and affection that inspires our bonds with the nonhuman natural world (1). Following the securities of religious faith, the anxieties of modernism, and the postmodernist confrontation with fragmentation and chaos, many nature writers explore new ways of belonging to the world, new ways of developing an ethic of caution and reciprocity in our interactions with nonhuman nature. Thus, a prime impulse of ecocriticism is to locate, open, and discuss this desire as it is expressed in cultural forms. SueEllen Campbell, in the final sentence of a lucid and inspiring essay on the relationship between ecocriticism and poststructuralist literary theory, epitomizes this critical impulse: "it is in nature writing—perhaps almost as much as in the wilderness itself—that I learn to recognize the shape and force of my own desire to be at home on the earth" (211).

The essays included in this volume display the range of ecological literary criticism from literary theory, to scholarly practice, to applied pedagogy. Several authors contribute to the discussion of what ecocriticism can accomplish and suggest means by which we might go about doing this work; most, however, analyze specific texts and offer thereby a richer understanding of the interplay of language and environment in literary works written throughout the past two centuries. Students, teachers, and scholars may learn from these studies how to recognize and examine the many important ways in which literary texts engage and express the complex relationships of human beings to the nonhuman natural world. In these twenty-one essays, readers will also see new kinds of meaning being born as authors work on the cultural borderlands where literature and environment meet. Collectively, these essays teach us how to read the environment in the literary text and how to read the literary text in the environment. They suggest new ways in which we might read books and, through books, how we might better learn to read the complex, extraordinary text that is the natural world.

While all the essays in this collection grew out of lectures and papers originally delivered at the first conference of the Association for the Study of Literature and Environment (ASLE), this volume does not represent the full range of energies, interests, abilities, and visions that met on the campus of Colorado State University in Fort Collins, Colorado, in June of 1995. We have not attempted only to select the "best" of the conference papers; instead, while seeking to develop a coherent structure for the collection, our goal has been to include a wide array of voices and topics that represent the vanguard of the discipline of ecocriticism. In an effort to adequately reflect these new directions in the study of literature and environment, we have divided the collection into four sections: Theoretical Perspectives on Culture and Environment; Genre, Gender, and the Body of Nature; Readings of Nineteenth-Century Environmental Literature; and Readings of Twentieth-Century Environmental Literature.

Scholars have occasionally taken ecocritics to task for their indifference to contemporary theoretical perspectives. However, the essays in Part I of this collection show that recent ecocritical approaches to literature frequently rely upon interdisciplinary models for textual analysis and speculate about the social and theoretical ramifications of literary expression. William Howarth's essay, "Ego or Eco Criticism? Looking for Common Ground," for instance, admonishes ecocritics to acquaint themselves with the physical sciences in order to offer legitimate interpretations of the ecological dimensions of environmental writing. Joni Adamson Clarke, in her piece called "Toward an Ecology of Justice: Transformative Ecological Theory and Practice," makes the case for a "transformative ecocriticism"—a critical strategy that employs multicultural perspectives in order to expose and thereby oppose oppression based on race and class. Still other authors included in Part I, such as John P. O'Grady, are working to reform the language of ecocriticism by eschewing scholarly apparatus and using personal narrative as a foundation for a mode of philosophical reflection that might reach a wider audience than do more conventional modes of literary criticism.

In contemporary literary criticism, explorations of both gender-inflected expression and representations of the body have challenged traditional modes of reading. Part II of this volume demonstrates that much ecocritical work has roots in these important fields of study. Such seminal works as Annette Kolodny's *The Lay of the Land: Metaphor as Experience and History in American Life and Letters* (1975) and Carolyn Merchant's *The Death of Nature: Women, Ecology, and the Scientific Revolution* (1980) have inspired ecocritics to study how language reveals dominant cultural attitudes toward the relationships between gender and nonhuman nature. We see examples of such work in J. Gerard Dollar's essay, "Misogyny in the American Eden: Abbey, Cather, and Maclean," which investigates the tension in texts that accept an "antithetical relationship between women and wilderness" but which also have narrators who challenge their culture's domination of the nonhuman world. Other scholars of literature and environment have been influenced by the theoretical work of ecofeminism, which takes as its focus the parallel oppressions of women and nature. In "The Body as Bioregion," for example, Deborah Slicer merges personal narrative and philosophical cultural criticism in order to condemn the ways in which social meanings define, control, and delimit a woman's experience of the bioregion that is her own body.

Most ecocritical scholarship focuses upon American literature of the twentieth century, yet an increasing number of literary scholars have recently offered new approaches to studying the environmental implications of pre-twentieth-century literature. Part III of this collection aims to show how several prominent nineteenth-century American authors, some of whom are already much discussed from non-ecocritical points of view, have recently been treated by scholars of literature and environment. Although such figures as John James Audubon, George Catlin, and Henry David Thoreau—all of whom are represented in this part of the book—have long been recognized as writers of the American landscape, major canonical authors such as Nathaniel Hawthorne, Ralph Waldo Emerson, and Walt Whitman have not. In "Nathaniel Hawthorne Had a Farm: Artists, Laborers, and Landscapes in *The Blithedale Romance*," Kelly M. Flynn explains how the relationship of physical landscape to human labor is essential to Hawthorne's artistic purpose in *The Blithedale Romance*. In "Agrarian Environmental Models in Ralph Waldo Emerson's 'Farming,'" Stephanie Sarver historicizes Emerson's environmental philosophy by reexamining it in light of nineteenth-century agricultural practices. Daniel J. Philippon's "'I only seek to put you in rapport': Message and Method in Walt Whitman's *Specimen Days*" opens a little-known work by one of America's major poets to unusually fruitful ecocritical scrutiny. The critical essays in this part of the collection demonstrate the value of recent scholarly efforts to examine the relationship of culture to nature that is dramatized in both canonical and noncanonical American literary works of the nineteenth century.

Finally, in Part IV, this volume offers a broad range of approaches to twentieth-century literature, both to explicitly environmental texts and to literary works that are less obviously responsive to the environment. As in Part III, the articles collected here include ecocritical studies of authors not known specifically for their environmental writing and, in other cases, applications of particularly original approaches to such familiar nature writers as Annie Dillard,

Terry Tempest Williams, and Richard K. Nelson. Among the more surprising and innovative essays featured in this section of the collection are Paula Willoquet-Maricondi's study of Aimé Césaire and Peter Greenaway, Dana Phillips's analysis of postmodern novelist Don DeLillo, and Len Scigaj's theoretically inflected approach to "agency and homology" in A. R. Ammons's *Garbage*. Twentieth-century environmental writing, both in the United States and abroad, is a vast and swiftly growing field, and the articles selected for this collection are intended to suggest the breadth and flexibility of approaches to such literature rather than to embody a definitive, prescriptive methodology.

In recent years ecocritics have celebrated the extraordinary vibrancy of environmental literature, and some have been so bold as to claim this as one of the richest—if not *the* richest—areas of contemporary American writing. Increasingly, readers both within and outside the academy have come to appreciate the intimate connections between word and world, art and nature, book and earth. We hope this collection of ecocritical essays will demonstrate promising new directions in the study of literature and environment and will suggest the importance and passion of this scholarly enterprise.

Works Cited

CAMPBELL, SUEELLEN. "The Land and Language of Desire: Where Deep Ecology and Post-Structuralism Meet." *Western American Literature* 24.3 (1989): 199–211. Rpt. in Glotfelty and Fromm 124–36.

GLOTFELTY, CHERYLL, AND HAROLD FROMM, EDS. *The Ecocriticism Reader: Landmarks in Literary Ecology*. Athens: U of Georgia P, 1996.

KOLODNY, ANNETTE. *The Lay of the Land: Metaphor as Experience and History in American Life and Letters*. Chapel Hill: U of North Carolina P, 1975.

LOPEZ, BARRY, ET AL. "Natural History: An Annotated Booklist." *Antæus* 57 (Autumn 1986): 283–97.

MERCHANT, CAROLYN. *The Death of Nature: Women, Ecology, and the Scientific Revolution*. San Francisco: Harper and Row, 1980.

ORR, DAVID W. *Ecological Literacy: Education and the Transition to a Postmodern World*. Albany: State U of New York P, 1992.

WILSON, E. O. *Biophilia*. Cambridge: Harvard UP, 1984.

WINKLER, KAREN J. "Inventing a New Field: The Study of Literature About the Environment." *Chronicle of Higher Education* (9 Aug. 1996): A8+.

PART I
THEORETICAL PERSPECTIVES ON CULTURE AND ENVIRONMENT

Ego or Eco Criticism?
Looking for Common Ground
WILLIAM HOWARTH

It was time to stop. After two exhausting hours, our committee had not yet agreed to include literature in the environmental studies curriculum. Margaret, who studies fossil ferns, summarized the view of science:
  "Nature resists stories. It likes facts. It *is* facts."
  "*Fact* is from *facere*, to do or make," I said.
  She sighed: "You always revert to language."
  I shrugged: "And maybe you hide in numbers."
  Among humanists these days, it's not much easier to hold a dialogue. Everyone has strong views about canons, methods of reading, and the social roles of scholars. There's an emphasis on difference that spawns new fields, like environmental literature, but also threatens them with schisms. Already a gap has opened between social ecology on the left and deep ecology on the right, both embracing ecology to discredit science. More ego- than ecocentric, these approaches are unlikely to build rapport with other disciplines.
  Humanists need methods of study that are critical, imaginative, yet also rational. For me, literary ecology is best deployed on texts that put our minds to land and help us write about reading nature. Our readings will represent the interaction of nature and culture by reflecting how places shape thought. They offer us a chance to find common ground, a space that's large enough to share at no cost to our diversity.

Willa Cather's <u>My Antonia</u> (1918) is rarely taught in today's EcoLit 101, because Cather was unaware of toxic waste and too cheerful about frontier settlement. Yet <u>her novel clearly describes how one learns to read a land</u>. At the outset her narrator, Jim Burden, recalls his journey as a ten year old from the mountains of Virginia to live with his grandparents in Nebraska. Both of Jim's parents are dead, and to a young orphan this western grassland is so vast, treeless, and wind swept that it only increases his loneliness.

Then one morning his grandmother takes him out to her garden at the bottom of a draw, a broad gully cut into the prairie by stream erosion. After chores, Jim asks to stay in the draw for a while alone. He leans against a pumpkin and remains very still. Up on the level, the wind sings and the grass waves, but down in this sheltered haven, it's warm and quiet. He waits and watches, neither wanting nor expecting much to happen. In that mood, he begins at last to feel at home—a sense of belonging that comes to him from this outdoor place. As he recalls:

> I was something that lay under the sun and felt it, like the pumpkins, and I did not want to be anything more. I was entirely happy. Perhaps we feel like that when we die and become a part of something entire, whether it is sun and air or goodness and knowledge. At any rate, that is happiness, to be dissolved into something complete and great. When it comes to one, it comes as naturally as sleep. (Cather 15)

In literary terms, this scene is memorable because Willa Cather has used a piece of her native American landscape to sustain a classical tradition, the <u>pastoral elegy</u>. In the elegy, natural scenes dissolve human grief. Down in the draw, Jim accepts his losses and gains, seeing death as a common destiny that makes him a part of something entire. Those sentiments have comforted many readers, including my grandmother. She grew up on prairie farms in Indiana and Illinois, and when she died at the age of ninety-nine, we read that passage from *My Antonia* at her funeral. Funerals are like that, moments when life wants to hear that death is no threat, only a passing away into something entire, a destiny as safe and natural as sleep.

Down in the draw, Jim's thoughts are dreamlike because this place, a hollow cup in the land, is so warm and sheltered that it provides him with a nurturing, womb-like security. Here all the natural elements—wind, water, earth, and light—flow into one place, which his grandmother has cultivated into a flourishing garden. In the passage, Willa Cather's prose is also highly cultivated. *Cultivate* means to raise up, to elevate in stature, an impulse that creates a problem for us as readers.

Although Jim says that he waits for nothing to happen, something soon does. As he turns from description to commentary his voice changes, slipping from a tentative phrase, "perhaps we feel like that when we die and become a part of something entire," to firm imperatives: "At any rate, that is happiness, to be dissolved into something complete and great. When it comes to one, it comes as naturally as sleep" (15). This moment is early in Jim's story and yet already he is directing us to see, feel, and think in particular ways. While it may be appropriate for him to brood about death at this time and place, we may also ask, why should he find that answer here and so soon?

Of course, the novel is an autobiographical story, told to us by an older, wiser Jim Burden. But in this passage it may be fair to say that Willa Cather stopped reading the land and began to write upon it. Knowledge no longer flows to Jim from the sun and earth; instead it rises from within or from a pastoral voice that speaks through him. If we are readers who admire large platitudes—"that is happiness, to be dissolved into something complete and great"—then we may follow this pastor to green pastures and still waters, there to lie down to sleep and be entirely happy. But if we are less sheepish, more resistant readers, we may well ask why that voice rises from this place, a prairie draw?

The draw has sloping sides, perhaps reminding Jim of his home in the Blue Ridge hills, yet he doesn't say so. The draw is also suggestive because it tells a history of loss and gain. Its slopes were cut by erosion and the bottom built through deposition, and those layers of damp, rich soil below are more productive than the dry, windswept plains above. Erosion and deposition are natural disturbances, events that constantly recur and sustain the dynamics of an ecosystem. Through them the draw forms and grows fertile, inducing Jim's grandmother to plant its soil and make it yield crops. Her actions follow an instinctive land ethic, built on observing nature and adapting to its ways. Jim may at first grasp that principle when he expects little to happen, and yet, as we've noted, his rhetoric soon makes large claims on our attention.

Cather wants to deliver a message through Jim, something that helps dissolve his loneliness into a feeling that is complete and great.

My sense is that she is using landscape in the conflicted ways that have long been typical of Western culture. Nature on the one hand is seen as something complete and great, vaster than our powers to know, but it also is a ground that we shrewdly exploit, often to its detriment. That garden down in the draw is an ironic sign of the prairie's future, for one day this empty, boundless region will bear houses, fences, and roads, the "improvements" that measure the progress of a westering nation.

Cather was endorsing that frontier ideology as late as 1918, even though the 1900 census revealed that population was declining in the rural West. In less than a century, settlement of the High Plains resulted in an environmental disaster. The destruction of native tribes, of bison, and of prairie soil and the construction of railways and towns were all early steps in a process of erasing the region's original ecosystem. Homesteaders who followed Jim's grandmother onto the Plains did not plant down in the draws but up on the levels. They tried to make Nebraska another Ohio or Illinois by converting a natural grazing range into cropland, replacing an intricate web of land and life forms with the monoculture of row crops, wheat and corn.

To plant they broke the ancient sod. They tore up its twelve-foot root systems and turned over the damp soil exposing it to sun and wind. At first the land gave good yields. But in decades its soil weakened, baked by drought and blown away on prevailing winds. From those years onward the region steadily declined. Today the American grasslands, heart of the continent, are a depressed and depopulated region. According to demographers, High Plains statistics—job losses, declines in population, education, and prosperity—closely match another blighted region, the urban inner city. Between Bedford-Stuyvesant, New York, and Hayes Center, Nebraska—250 miles west of Cather's Red Cloud—lies far too much common ground.

In literary work we always face two dangers, reading too much into a text or reading too little. If this moment with Jim Burden serves for construction as a parable, it suggests several implications about change and growth, the usual course of development. The environmental humanities presently face the new dilemma of sudden popularity. At literature and the environment conferences these days, publishers are seeking authors and backpacks outnumber tweed jackets. After years as a self-styled loner, I'm meeting a great many like-minded people.

We are caught between the draw and the levels, the spaces of solitude and of community. Like Henry Thoreau, we see the outdoors as our studies, not an escape, and that attitude no longer seems eccentric. Living in an era of prolonged environmental crisis, we find that examining natural-cultural relations has at last become a trend. For us the frontier days are over. Trains are arriving and they bear homesteaders looking for cheap land. The question rises: what should they settle? Between the draw and the levels, where lies a common ground?

Academics often long for power, but the main goal of environmental study is health. Health comes from what native people call medicine, a force that restores to culture its sense of well-being. In that spirit, let me offer three pieces of medicinal advice.

First, the health of environmental criticism lies not in numbers, but in variety. Growth will come, but if seen as an end in itself, it may produce an intellectual monoculture. Becoming too like-minded will harm our effectiveness. If we have different attitudes and outlooks about our work as scholar-teachers, if we have differing agendas in research and publishing, we'll be going about our evolution in a healthy way.

Second, health lies in knowing more disciplines than literature. If we call ourselves ecocritics, we should know ecology and its place in the history of earth sciences. I'm not a professional scientist, but to read that passage in *My Antonia* I used ecology, geology, and biology and asked how they shaped history. We need to understand evolution and genetics, speciation and habitats, the human genome project or any other scientific activity, in order to make responsible statements about environmental issues.

Third, health lies in achieving a balance between work and play. The study of environmental literature is a vocation, not a vacation. Many of our conferences have long lists of outdoor activities, understandably so. After weeks of professing, the season beckons and it's time to climb slopes or ride white waters. But our meetings should also feature serious workshops on editing and writing, where scholars can work on building research agendas.

The goal of this health plan is broader professional recognition. We will survive only if we build a tradition of distinguished scholarship. In the history of the academy, intellectual fields are established by attracting research scholars. They draw students to graduate programs, which provide rigorous training. The training should clarify how nature and literature interact in ways that appreciate but also analyze language with sharp critical tools. The courses must have genuine content in order to be truly interdisciplinary—and satisfy colleagues in nonhumanities fields.

Currently I teach a large undergraduate course on literature and environment that draws half its students from the humanities and half from the social or natural sciences. The course brings together poets and engineers who read Annie Dillard or Norman Maclean, and many say, yes, this is an education! Our research will sustain teaching by offering students ideas and values about issues of real concern. This strong ethical component gives students a way to enhance their life's work—in policy, law, medicine, education, media, even business. That aim may sound conventional, but in today's climate it's actually unorthodox for humanists to deliver much applied value.

Administrators today regard the humanities as loss leaders, costly departments that produce little revenue. But the humanistic values of reason and empathy can focus on environmental problems, which arise from errors in human belief and judgment. The scientist's task is to predict, the humanist's is to remember. To remember with truth and compassion is to know the past and take steps toward a viable future. We need to convey that value to colleagues in science and thus humanize them. We can't huddle with Jim down in a warm, safe draw or plow the wind-swept levels. Ruined land can be restored, for it is alive—and it is a common ground where we dwell.

Work Cited

CATHER, WILLA. *My Antonia*. New York: Houghton Mifflin, 1954.

Toward an Ecology of Justice
Transformative Ecological Theory and Practice

JONI ADAMSON CLARKE

The ethic of reconciliation with the earth has yet to break out of its snug corners of affluence and find meaningful cohesion with the revolution of insurgent people.

ROSMARY RADFORD RUETHER, "MOTHEREARTH AND THE MEGAMACHINE" (51)

On New Year's Day 1994, Mayan insurgents calling themselves the Zapatista National Liberation Army burst out of the great Lancandon rain forest and took over the colonial town of San Cristobal de la Casas in the state of Chiapas, Mexico. Politicians, journalists, and scholars are still scrambling to make sense of these dramatic events and the ongoing peace negotiations between the Zapatistas and the Mexican government. But as news of the rebels's demands are reported in the international press or distributed worldwide over the Internet, it becomes increasingly clear that this developing revolutionary movement is more than a simple peasant uprising.

When Subcommandante Marcos, one of the masked leaders of the Chiapas revolt, was asked by journalist Ann Bardach, "[W]hat is behind this revolution in Mexico?" he answered that the rebels want their land back (73). Members of the Zapatista National Liberation Army are predominantly Mayan Indians, so Marcos's statement is not simply a reference to the fight to preserve pristine rain forest ecosystems; rather, he is referring to the Mayan Indians's centuries-old fight to save lands which have been the basis of their cultural and economic survival for thousands of years. The rebellion is taking place on the borders of a rain forest which the Mexican government has repeatedly pledged to protect. But in the last few decades, the government has looked the other way while the Lancandon has been shorn for timber, highways, farms, oil drilling, resettlement, and even airstrips for drug traffickers. With the expropriation of lands that were once cultivated with traditional crops and with the disappearance of the plant and animal species that once supported them, the Mayans are forced to abandon their villages to work as wage laborers in the countryside or seek work in the cities, where they usually live in poverty and are able to practice Mayan traditions only with great difficulty.

At the initial round of peace negotiations in February 1994, Marcos made it absolutely clear that the Zapatistas see political and cultural reforms as inextricably connected to land reforms and environmental protection. The rebels demanded an end to the human rights abuses that indigenous peoples have suffered for centuries. They also demanded election reforms which will ensure that Mexico's indigenous peoples will have a voice in shaping the country's future. From the perspective of the Mayans who have donned the mask of rebellion, any discussion about the preservation and conservation of the Lancandon rain forest must also address the basic human rights of the peoples who call the rain forest home.

In *Literature, Nature, and Other*, Patrick Murphy observes that there has been a recent surge in the numbers of scholars and teachers interested in the field of environmental literature and ecocriticism (165), and notes that most discussions of ecocriticism to date have not adequately treated the issues of race. In the discussion that follows, I would like to build on the perspectives that are being articulated by those involved in environmental justice movements—like that of the Zapatistas—to argue that the issues of race and human rights must be brought into any satisfactory ecocritical discussion of "nature" and/or "nature writing." I will focus on the work of two Native American authors—Acoma Pueblo writer Simon Ortiz and Laguna Pueblo writer Leslie Silko. Both writers are raising many of the same issues in their poetry and fiction that the Zapatista rebels are raising. I will argue that we must bring the perspectives of such multicultural writers into our study and teaching of literature if we would practice what I am calling a "transformative ecocriticism."

In his chapter "Voicing Another Nature," Patrick Murphy observes that several recent anthologies of "nature writing" have characterized the form as factual prose essays which describe in a lyrical and detailed way some pristine area untouched by humans. He adds that Native American literature has often been excluded from such anthologies because it "does not admit a separation between facts and fictions" and because it does "not easily admit a clearcut division between prose and poetry" (Murphy 44). However, the very ways in which Native American literature defies the traditional generic conventions of "nature writing" raise some important issues which should be considered by those interested in reading, studying, and teaching in the field of environmental literature. It has become a commonplace to note that contemporary American Indian writers examine the relationship between humans and the land, but what interests me is that Native American poetry and fiction are almost never set in "pristine wilderness" areas "untouched" by humans. Rather, this literature is usually set in places that make post-Columbian colonization of American lands obvious: inside the contested boundaries of a Central American rain forest or a North American Indian reservation, on the U.S./Mexico border, within the dangerously precarious borders of a U.S. military testing and bombing range, or behind the barbed fences of a multinational corporation's uranium mine.

For example, in his poem, "That's the Place the Indians Talk About," Simon Ortiz writes about Coso Hot Springs, which is considered a sacred, healing place by native Shoshonean peoples. A Paiute elder tells the narrator of the poem that

> When you pray.
> When you sing.
> When you talk to the hot springs.
> You talk with it when it talks to you.
> Something from there,
> from down in there is talking to you.
> You could hear it.
> You listen.
>   Listen. . . .
> Something is doing that
> and the People know that.
> They have to keep talking.
> Praying, that's the Indian way. (322)

The sacred hot springs are now fenced within the China Lake Naval Station, a center for the development and testing of U.S. military weapons. Now the people must "talk with the Navy people" to get permission to "talk with the hot springs power" (Ortiz 323). But year after year, the Paiute elder and his people keep returning to the hot springs to hear voices bubbling up from deep within the earth and to listen to "the moving power of the voice, / the moving power of the earth, / the moving power of the People" (Ortiz 324). Year after year, their return is an act demonstrating that the fight for the land is not simply a fight for pristine wilderness or endangered species; it is a fight for home, a fight for the land that has been the basis of Shoshonean spiritual, cultural, and economic survival for hundreds of years. The fence which separates Shoshonean peoples from their sacred springs becomes a metaphor for the ways in which the oppressions of people and land are linked. Though it does not conform to traditional generic conventions, I would argue that Ortiz's poem is "nature writing" that theorizes that if we would fight for the land, we must explore the connections between the oppression of certain races and classes and the appropriation and exploitation of indigenous lands.

Ecofeminist Karen Warren has explained that the connections between the oppression of nature and "Others" (women, people of non-European races, and people of the underclasses) are ultimately "*conceptual*: they are embedded in a patriarchal conceptual framework and reflect a logic of domination which functions to explain, justify, and maintain the subordination of both" (7). By examining the connections between the appropriation and exploitation of indigenous lands and the oppression of indigenous peoples, Simon Ortiz's poetry reveals and confronts the logic of domination which undergirds both. The Paiute elder and his people illustrate how the appropriation of traditional lands violates basic human rights because it makes the continued practice of indigenous lifeways extremely difficult or impossible. This is "nature writing," then, that makes the connection between environmental protection and social justice clear.

Leslie Silko's *Almanac of the Dead* presents an even more daunting challenge to those who insist on traditional definitions of "nature writing." This is not a novel which lyrically describes some "pristine wilderness area." Moreover, Silko's novel does not admit a separation between fiction and fact since it sets fictional characters amidst the numerous uranium mines and tailings piles, which are a reality in the American Southwest. In one of the novel's numerous, complexly overlapping story lines, Sterling, an older Laguna man, travels to the

Jackpile mine near the Laguna Pueblo. This massive, open-pit uranium mine is the same "point of convergence" where Tayo, the main protagonist of *Ceremony*, finally understands "a convergence of patterns."[1] Sterling has come to this "point of convergence" to visit a sacred sandstone snake which has recently emerged from a uranium tailings pile at the mine. Crawling through a barbed-wire fence that marks the boundaries of a gaping crater left by the abandoned uranium mine, Sterling sees thirty-foot mounds of virulently radioactive slag—uranium tailings which blow in breezes that carry them to the springs and to the Rio Paguate. "Here was the new work of the Destroyers," Sterling thinks, "here was destruction and poison. Here was where life ended" (*Almanac* 760). But, remarkably, it had been here, in this environmentally exploited place, that the ancient sandstone snake had emerged.

By surfacing in an area of the American Southwest that had become so radioactively contaminated by 1972 that the Nixon administration sought to have it designated a "National Sacrifice Area," Silko's snake calls attention to a profound irony. The Pueblo people's creation myths tell them that their ancestors migrated up through underground worlds to emerge from the *sipapo*, the sacred place from which all life emerges. Ironically, the U.S. government selects this as the place in which mines are sunk deep into the earth to extract the very substance that could ultimately destroy all life. Sterling recalls that Laguna elders had cried when the U.S. government opened the mine. "Leave our Mother Earth alone," the old folks warned, "otherwise terrible things will happen to us all" (*Almanac* 759). Nevertheless, government scientists and developers claimed a mandate to save humankind and proceed with a project that ultimately led to the deaths of many Navajo and Pueblo peoples and the nuclear contamination of their lands. Later, multinational corporations, like giant Anaconda, took over the mining operations for the government. But in the 1980s, when the uranium market bottomed out, these corporations abruptly ceased operations. The Laguna people, however, could not simply move away. They had no choice but to breathe the dusty, windblown tailings which today retain up to eighty-five percent of the original ore's radioactivity (LaDuke and Churchill 126).

By having Sterling transgress the fence around a multinational corporation's uranium mine, Silko calls attention to the ways in which modern states and multinational corporations claim a mandate to destroy lands and inflict suffering on millions of people in the name of scientific progress and development. The fence in this passage, like the fence in Ortiz's poem, becomes a metaphor for the ways in which the oppressions of lands and peoples are linked. When Sterling reaches the mine's gaping crater (the same "point of convergence" where Tayo finally understands a "monstrous design" of destruction and death), he also finally understands that lamenting the passing of a once beautiful environment is not enough. To effectively fight for the land, one must understand how a logic of domination undergirds the linked oppressions of certain races and lands.

Because Native American literature insists that the fight for the environment, properly understood, is the fight to end *all* forms of linked oppression, it seems to me that Native American writers—like the Zapatista rebels—are calling for something very similar to what Susanna Hecht and Alexander Cockburn term an "ecology of justice." In *The Fate of the Forest*, Hecht and Cockburn focus on indigenous activism in the Amazon rain forest, explaining that those working toward an "ecology of justice" are not working to "save nature." Rather, they are working for a "social nature" where humans and nature are not seen as separate entities and where the practice of justice restructures the concepts of nature (196–207). Hecht and Cockburn illustrate their point with examples from the Amazon rain forest, but their argument would have been equally convincing had they focused, as Ortiz and Silko do, on the enclosure of sacred places inside military installations or on the nuclear colonization of the Four Corners area in the American Southwest. By setting their poetry and fiction in such contested places, contemporary Native American writers proclaim to the politician or chief executive officer that land is not real estate or a space over which people can be moved like objects. Land is memory, a map of one's world, and the basis of a community's cultural and economic survival (Visvanathan 54). As J. Baird Callicott has so eloquently put it, this sort of nature writing acknowledges that our recognition of immersion in the biotic community "does not imply that we do not also remain members of the human community . . . or that we are relieved of the attendant and correlative moral responsibilities of that membership, among them to respect universal human rights and uphold the principles of individual human worth and dignity" (93).

No Native American writer that I know of has explicitly speculated about how we might reach the elusive destination of an ecologically just society. However, in *Almanac of the Dead*, Leslie Silko offers some clues. She writes about a Holistic Healers Convention attended by men and women of all races, classes, cultures, and backgrounds. Although a wide range of contradictory political views are represented at this convention, many of the participants wish to work for a more balanced, less violent world. These participants do not come together with a sense of ecology as managerial science. They recognize that the governmental or corporate systems which have claimed a mandate to set national borders, determine reservation boundaries, erect fences around gaping mines, or initiate massive development projects that displace thousands or hundreds of thousands of people, cannot be the basis of an ecology of justice. Rather, they come together with a sense of ecology as *communitas*. They are working to develop a better sense of the ways in which traditional tribal communities, like the Acoma and Laguna, recognized their responsibilities toward and connections to each other, to the total community, to nonhuman species, and to the land. The community ethic is based not on what each individual can take, but on what each can give. Silko does not depict the outcome of the Holistic Healers Convention and does not speculate about whether or not an ecologically just society is even possible, but the novel implies that an understanding of how the oppressions of race and class are connected to the oppression of nature is a requisite step toward ending the violence and domination preventing us from working toward an ecology of justice.

When we incorporate multicultural perspectives such as those of Ortiz and Silko into our discussions, we do not simply broaden or redefine the genre of "nature writing." Rather, we open our discussions to diverse perspectives which allow our understanding of environmental literature and ecocriticism to become richer, better, and more complex. To follow SueEllen Campbell's argument about the ways in which differing literary theories play off one another, the differences in each of these perspectives pinpoint the blind spots and reveal the possibilities in the others. I think this is the point that Barry Lopez is making in his discussion of American nature writing in *The Rediscovery of North America*. Lopez acknowledges the invaluable contributions to natural history made by early Euro-American essayists such as John Bartram, Thomas Nuttall, and John James Audubon. Then, employing indigenous perspectives, he pinpoints a blind spot in Eurocentric descriptions of the natural world: "[Eurocentric writers] were able at least to describe what they found.

But this extensive knowledge was ultimately regarded as only a kind of entertainment. Decorative information. A series of puzzles for science to elucidate. It was never taken to be what it in fact is—a description of home" (31).

In Simon Ortiz's and Leslie Silko's work, the land is not depicted as a puzzle for science to elucidate but is depicted instead as home. There is no lamentation over a fast-disappearing wilderness or the last Indian warrior. Rather, Ortiz's Paiute elder and Silko's character Sterling insist that ancient American Indian cultures, though transformed by the conflagration, survive and continue despite conquest and colonization. As an act of resistance to the government's unjust domination of one of his people's sacred places, the Paiute elder keeps returning to Coso Hot Springs. Despite radioactive contamination, Sterling's return insists that his people will continue to fight for and inhabit their ancestral homeland. Like the Zapatista demands laid on the negotiation table in Mexico City, then, the poetry and fiction of many contemporary Native American writers make us aware of how the fight to protect the environment is connected to the fight for basic human rights.

Karen Warren has argued that a "transformative feminism" would address the conceptual and structural interconnections among all forms of domination. In this way, it would encourage feminists concerned with ecology to join allegiance with those seeking to end oppression by race and class. Otherwise, feminist concerns over ecology would degenerate into a largely white middle-class movement (19). Following the argument of Rosemary Radford Ruether, Warren adds, "The promise of transformative feminism requires making connections with 'the revolution of insurgent peoples'" (Warren 19).

Building upon the work of Warren and Ruether, I would like to argue that a "transformative ecocriticism" must address the conceptual and structural interconnections among all forms of domination. Incorporating multicultural perspectives and literatures into their study and practice would encourage critics and teachers concerned with ecology to join allegiance with those seeking to end oppression by race and class. Only by recognizing the connections between the oppressions of certain peoples and places can literary critics begin to develop ecological theories and practice that can work transformatively toward an ecology of justice.

Note

1. *Ceremony* (246, 254). The tailings piles outside the Laguna Pueblo that Silko describes in *Ceremony* are located at the now abandoned Jackpile uranium mine which opened in 1952 and which was once the largest in the world. For more about the effects that uranium mining has had on Pueblo and Navajo people, see Ortiz (22, 354–56).

Works Cited

BARDACH, ANN LOUISE. "Mexico's Poet Rebel." *Vanity Fair* July 1994: 68–74, 130–35.

CALLICOTT, J. BAIRD. *In Defense of the Land Ethic: Essays in Environmental Philosophy.* New York: SUNY P, 1989.

CAMPBELL, SUEELLEN. "The Land and Language of Desire: Where Deep Ecology and Post-Structuralism Meet." *Western American Literature* 24.3 (1989): 199–211.

HECHT, SUSANNA, AND ALEXANDER COCKBURN. *The Fate of the Forest: Developers, Destroyers, and Defenders of the Amazon.* New York: Verso, 1989.

LADUKE, WINONA, AND WARD CHURCHILL. "Native America: The Political Economy of Radioactive Colonialism." *The Journal of Ethnic Studies* 13.3 (Fall 1985): 107–32.

LOPEZ, BARRY. *The Rediscovery of North America.* Lexington: U of Kentucky P, 1990; New York: Vintage, 1992.

MURPHY, PATRICK D. *Literature, Nature, and Other: Ecofeminist Critiques.* New York: SUNY P, 1995.

ORTIZ, SIMON. *Woven Stone.* Tucson: U of Arizona P, 1992.

RUETHER, ROSEMARY RADFORD. "Motherearth and the Megamachine: A Theology of Liberation in a Feminine, Somatic and Ecological Perspective." ED. Carol Christ and Judith Plaskow. *Womanspirit Rising: A Feminist Reader in Religion.* San Francisco: Harper and Row, 1979. 43–52.

SILKO, LESLIE MARMON. *Almanac of the Dead.* New York: Simon and Schuster, 1991.

———. *Ceremony.* New York: Penguin, 1977.

VISVANATHAN, SHIVE. "From the Annals of the Laboratory State." *Alternatives* 12 (1987): 37–59.

WARREN, KAREN J. "Feminism and Ecology: Making Connections." *Environmental Ethics* 9.1 (Spring 1987): 3–44.

Talking about Trees in Stumptown
Pedgogical Problems in Teaching EcoComp

MICHAEL MCDOWELL

Environmental issues make ideal subjects for composition classes because they are as complex, as multidisciplinary, and as emotionally charged as any social issues can be: they are based upon cultural assumptions that are currently changing; every student has direct personal experience with them; and many environmental issues engage every sense we have. As more faculty merge ecocriticism and environmental literature in their freshman composition courses, it's becoming apparent that several problems commonly arise in an EcoComp classroom that don't typically arise in other kinds of composition classes. An environmentally oriented class can become an alienating experience for some students and can lead to frustratingly simplistic thinking and writing if basic problems common to EcoComp aren't addressed.[1]

The first and hardest problem to overcome is the effect of the popular media's definition of environmental issues. Most of the media dwell on massive global issues, such as global warming, acid rain, and population growth. In general, the media also tend to polarize the issues they recognize. The focus is usually on institutions and spokespeople's responses to recent events, instead of on the underlying questions of how a culture relates to its landscape and how cultural assumptions determine those relationships. So students come to class, and sometimes continue throughout the term, believing that the real environmental concerns are whether or not to recycle, to limit pollution, or to preserve old growth, wilderness, the ozone layer, whales, or tropical rain forests—usually elements of the natural world they never come into contact with. Often the underlying structure of their essays is a simple polarization of jobs versus the environment, reality versus idealism, present versus future, or preservation of the status quo versus extreme immediate change. When environmental issues are defined as a choice between polar opposites, nearly everyone chooses the seemingly safer, more familiar, more certain, and more comfortable nonenvironmental option. A first lesson, then, is that ready-made, media-simplified issues should be off-limits, since whoever defines an issue typically carries the day in argument.

The recent appearance of environmentally oriented texts for composition courses has considerably alleviated this problem of definition. These texts—notably Morgan and Okerstrom's *The Endangered Earth* (1992), Slovic and Dixon's *Being in the World* (1993), Levy and Hallowell's *Green Perspectives* (1994), Walker's *Reading the Environment* (1994), Anderson and Runciman's *A Forest of Voices* (1995), Ross's *Writing Nature* (1995), Valenti's *Reading the Landscape* (1996), and Jenseth and Lotto's *Constructing Nature* (1996)—are generally structured to emphasize humans' responses to the environment, instead of emphasizing the kinds of issues the media focus on. These texts also present many personal essays in which writers discuss their own responses to unique, specific places. But the national status of the publishers and the economies of scale in which the publishing companies operate require that they appeal to more than a regional market, which creates a second problem in the EcoComp classroom. What these texts can't do is direct students through their own regional histories and issues and their own regions's literary responses to the environment. Students generally must respond either to readings on global topics or to readings about places and issues which aren't their own and which seldom lead them closer to a sense of their own place.[2]

The most appropriate solution to this problem is suggested in the writing assignments at the ends of chapters in the texts: students should adopt the approach taken by the article just read and research their own locale for similar issues. At this point we confront a third problem in teaching EcoComp: a dearth of easily accessible information about local environmental issues. In the first two terms of a freshman composition sequence there usually isn't time for adequate research of complex environmental issues. More than half my students do unrequired research for every essay, but too often what they find in the library is national or global and doesn't bring them back to local specifics. (In the research-oriented third course of the composition sequence, students have succeeded far better with environmental issues.) I also hand out lists of local environmental organizations which lead students with cars, adequate time, and initiative to excellent information.

My college is now experimenting with FIGs, or Freshman Interest Groups, which can help students gain a substantial information base for discussing environmental issues. The idea is to have students register to take a cluster of courses offered by different departments and taught by instructors who have complementary goals for the term. In the first term the college offered FIGs, for instance, I taught an environmentally oriented composition class in which two-thirds of the students were simultaneously taking a cultural geography class focused on environmental issues and a political science class focused on the environment and peace and conflict studies. I could count on at least two-thirds of the students being currently engaged outside of our class in pertinent discussions bearing directly on our topics, but with depths and perspectives I'm unable to provide.

Before FIGs, and before this wealth of environmentally oriented composition texts appeared, I composed packets of readings about issues relevant to our particular locale. We'd spend two weeks reading articles about Northwest wild salmon issues, and another two weeks reading about Northwest old-growth forest issues, Northwest wilderness, wetlands, or Oregon land-use issues. I stopped using the packets, though, because of the immense time involved in updating and securing copyright permissions for several hundred constantly changing articles.

Eventually I began to require our local daily newspaper, *The Oregonian*, as a text. While hardly proenvironment, the newspaper does have several writers devoted to environmental issues. The result was heartening: students who typically never read daily newspapers, or even magazines, became informed about and engaged with developments in various local issues, but only occasionally did they gain much insight into any particular issue by relying only on a newspaper. Also, the preparation time for the class expanded drastically; six or eight often lengthy articles on environmental topics in every day's paper called for three times the preparation that teaching from a textbook did.

Although now I only recommend a daily newspaper, the inclusion of daily environmental news into the class has expanded the definition of EcoComp to embrace almost every subject a student can think of. Whereas initially I thought of these courses as focusing on the natural world, now they focus on the environmental aspect of any significant issue, the closer to home the better. We discuss a typical day, looking at the environmental implications of every small action, such as flushing a toilet, breakfasting in midwinter on an orange and banana, and inhaling formaldehyde from pressboard furniture and wall-to-wall carpeting. We discuss land-use planning, neighborhood covenants, urban green spaces, sewage overflows, light-rail placement, projected local population growth, and local business practices. Nature and environmental issues are not something "out there" beyond the limits of the metropolitan area's million-plus people, but are everywhere in the cities we inhabit. Only those with some personal experience with wilderness or logging or livestock grazing tend to write about purely rural subjects.

It would be good if established publishing companies would produce regionally focused textbooks. Perhaps the solution is for the next generation of EcoComp textbooks to have regional supplements, as Melissa Walker has suggested. At the very least, end-of-chapter or end-of-unit questions and suggestions for writing could be organized regionally, focusing attention on the most pertinent aspects of the issues for each of the United States's nine or ten major regions. Or maybe changing the textbooks is the wrong approach, since after all we do need to understand the larger national and global pictures; perhaps we should work instead to establish a clearinghouse of regionally specific, eminently "teachable" environmentally oriented essays available quickly and inexpensively, perhaps through the World Wide Web.

As long as the student finds an issue which thoroughly captures his or her attention and about which the student can be specific, I don't really care what a student writes about or what position the student takes. Those early-in-the-term desires to write about recycling and pollution tend to disappear after discussion. Recycling is a fine and necessary activity, we typically agree, but it also is a tactic to foster quietism by distracting our attention. Pollution mitigation becomes less important as a topic when we look at every environmental issue as a health issue, and when we ask such questions as, "How much cancer is okay in *my* body?" Typically we agree that it's better to address causes than symptoms. Part of learning to write an essay is learning to question assumptions and to understand the reasoning of those who might oppose the writer's position. I ask for a fair discussion of the best reasons for informed and intelligent opposing points of view. I also ask for a demonstration that the writer has understood and respects those people of goodwill who oppose the writer's position; for example, we discuss tunnel vision as a sign of failure in an essay.

Unfortunately, it's usually easier for students to question the assumptions of those with whom they disagree than it is for them to question their own assumptions. There are several culture-bound beliefs basic to mainstream American thought which almost no one in a typical class will allow to have questioned, which presents a fourth major problem. One not-to-be-questioned assumption is that private property is sacrosanct and that no government agency can dictate use to a property owner. Another unquestionable assumption is that property can be owned in the first place. Another is that every human being has the right to have as many children as they want. No matter how much energy and enthusiasm has been put into a discussion, when these assumptions are held up for analysis, the discussion up to that point is forgotten, and typically the entire class expresses outrage that there could be serious opposition to the idea of owning property or having the freedom to make another child. The resistance to questioning these assumptions is so amazingly great and the benefit to the discussion of our principal subject often so minor that I usually just point to other cultures which haven't held these assumptions to show that these are choices, not immutable laws, and we move on.

A fifth major problem in teaching EcoComp is the tendency for both students and teachers to try to persuade others to agree with their positions; this question of advocacy in the classroom never goes away. Six years ago when I began teaching EcoComp, everyone whose father or uncle worked in the woods and everyone from timber communities on the outskirts of the metropolitan area dropped the course within the first three weeks, robbing the class of balance. I didn't want the classes to become groups of like-thinking people, as they would have except for the fact that so many students said they didn't know what to think at all about environmental issues. Their preoccupation with jobs, family, and schoolwork and the community college's implicit promise to lead them to better jobs tended to discourage lofty thinking about larger issues.

In the last four years, most students agree that everyone needs to think about environmental issues. The discussions have improved. And every class has one or two students who believe in and can articulate the proindustrial, prodevelopment ideology on environmental matters, and they stay with the class. We hear, in the midst of environmental discussions, the commonplaces that American business is great, progress is undeniable, everyone pursues only his or her own self-interests, everyone is greedy, the most advanced technology is always the best technology, and all problems can be solved by technology. Oddly enough, at students's request I often find myself explaining these popularly held beliefs more than environmental stances. In the process I sometimes sound like an advocate of traditional American business.

The question of fairly representing the stances of business and industry arises constantly. At first I thought it unnecessary to explain business or industrial positions because these positions were evident everywhere: we see them in their concrete, steel, and asphalt versions in our landscape, we hear them in endless advertisements and discussions of the economy, we inhale them with every breath of postindustrial air. Yet students continually ask for readings and discussion on what a business would think about any particular issue. They want to hear other points of view specifically validated. And while my goal is to include as many points of view as possible before students take a stance, it's difficult to find articulate, thoughtful explanations of environmental positions taken by businesses. Most statements from businesses beg the questions, present irrelevant facts and statistics, and appeal to emotion more than to logic. People clear-cutting a hillside to build half-million-dollar houses don't write essays probing the meaning of what they're doing. For this reason, arguments presented by the "wise use" movement have proven invaluable as voices against the more familiar environmentally friendly voices. Wallace Kaufman's essay "Confessions of a Developer" from *Orion* magazine's *Finding Home* is a favorite of all students. Another EcoComp text that does a superior job of playing off business and environmental positions is Carol Verburg's *The Environmental Predicament*.

In addition, I've learned the value of employing a kind of parenthetical citation during discussions so that ideas are attributed to their sources. To present multiple points of view, I try to attach an author to each opinion and disappear myself, so it's the Wilderness Society or the Audubon Society or the BLM or the Forest Service or Weyerhaeuser or the Inter-Tribal Fish Commission who say this, not me. I'm interested in recreating the dynamics of the major players in the issues, not in offering my own synthesis as the "correct" or intelligent stance. I "model" argumentative methods often enough as it is, and my goal in class discussions is to keep positions fluid for students to work with in their own essays.

These oral parenthetical citations help to decenter the classroom, removing the instructor from the center stage and from the artificial role as font of truth and sole dispenser of sweetness and light. Such decentering isn't hard after adopting a method based upon the ecological sciences' idea of all physical reality being a complex web of relationships. In argument-based composition classes, the usual textbook metaphors for argument too often derive from war and the military: we take a position, define its perimeter, and then focus and launch an attack on our opponent, marshaling our forces and eventually bringing in the big artillery, and hopefully "winning" the argument at the end. Or textbooks present a courtroom model, with two opponents who, like plaintiff and defendant, find fallacies in each other's arguments until both sides can agree, in a sort of pragmatic, compromising settlement. The martial and judicial models of argument present my sixth and last problem in teaching EcoComp: I believe that other metaphors better explain what we're attempting to do in our writing. For instance, argument takes on a new tone when looked at as play: the writer moves around the playground from one position to another. The writer may not exactly have a position at the start, but that's okay: the goal is not to "win" but to try out new perspectives, to inquire into possibilities, and to discover or arrive at truths for our time and our place.

Better than "play" is an ecological metaphor: the writer interacts with other elements of the intellectual or social ecosystem. And, as in all ecosystems, the interaction with other elements changes the writer; only in the interaction with others does the writer define himself or herself. And the diversity of the others with whom the writer interacts determines the complexity and richness of the writer's new (and temporary) self-definition. The critic who most exactly explains such an ecological conception of rhetoric (without ever calling it ecological), Mikhail Bakhtin, explains that the others' kinds of language, professional and ethnic backgrounds, social status, and age, contribute to the writer's process (288–91, 314). Seen ecologically, the discourse community created in the classroom becomes critical to the writer's well-being; without those other voices to participate in a dialogue, the writer has little hope of modifying positions, adapting to others' needs and demands, and continuing to survive.

Such a way of looking at essay writing leads to far better essays. Rather than student essays that shout, "I'm right! Everyone else is wrong, and here's why," students encounter the reasons, not just the positions, of those with whom they disagree, and often they find grounds for agreement where they'd least expect. An initial worry of how to handle cranks in the classroom has evaporated with an ecological approach. The attitudes of "Death to loggers" and "All businesses are antienvironmental" become impossible to hold while interacting with opposing sides. To ensure that interaction, I stipulate that a student writer hasn't adequately analyzed a problem until he or she has understood and sympathetically represented the point of view of every major party of goodwill involved in the problem. If we persuade a writer to consider each of five or six perspectives on an environmental issue, we've succeeded in making EcoComp the kind of course that stimulates a student's intellectual growth.

The problems arising from teaching EcoComp are small compared to the benefits of an EcoComp class for the students. Students quickly find themselves grappling with issues that directly affect their lives and all of a sudden the arguments under discussion aren't simply theoretical or academic. Students also learn to accept and work with uncertainties in the midst of our culture's paradigm shift in our conceptions of the human relationships to the landscape. And because few issues are as complex and interconnected as environmental issues, students in an EcoComp class must learn to employ interdisciplinary approaches—a valuable experience for students who are continually under pressure to limit their viewpoint to that of a single discipline in which they'll major.

The students are only the most obvious beneficiaries. An EcoComp instructor knows that with each pencil mark on a student essay and with each well-conducted discussion, we've made it likelier that all members of the natural world—humans, animals, plants, rocks, oceans, winds—will live healthier and happier lives on a vigorously healthy earth.

Notes

1. EcoComp has quickly become the term of choice for those involved in teaching Ecological Composition, which might be defined as any freshman English composition course which uses as the principal texts an assortment of articles, essays, poems, stories, and books about nature, landscape, "the simple life," or issues arising from how humans have related to the natural world.
2. While on the one hand the publishing industry is becoming monolithic, with fewer and fewer large traditional publishing houses, by all accounts it is at the same time decentralizing, and the smaller, widely flung publishers are more sensitive to their own regions. Such a fragmenting encourages ideas like Melissa Walker's suggestion to publish a text with supplements localized for different regions of the country. As an example of what we might expect from a decentralized publishing industry, Planet Drum Foundation of San Francisco has recently come forth with Peter Berg's *Discovering Your Life-Place: A First Bioregional Workbook*, which guides readers to an understanding of their own regions.

Works Cited

ANDERSON, CHRIS, AND LEX RUNCIMAN. *A Forest of Voices: Reading and Writing the Environment*. Mountain View: Mayfield, 1995.

BAKHTIN, M. M. *The Dialogueic Imagination: Four Essays*. Ed. Michael Holquist. Trans. Caryl Emerson and Michael Holquist. Slavic Ser. 1. Austin: U of Texas P, 1981.

BERG, PETER. *Discovering Your Life-Place: A First Bioregional Workbook*. San Francisco: Planet Drum, 1996.

JENSETH, RICHARD, AND EDWARD E. LOTTO. *Constructing Nature: Readings from the American Experience*. Upper Saddle River, NJ: Blair-Prentice Hall, 1996.

KAUFMAN, WALLACE. "Confessions of a Developer." *Finding Home: Writing on Nature and Culture from* Orion Magazine. Ed. Peter Sauer. Boston: Beacon, 1992. 38–55.

LEVY, WALTER, AND CHRISTOPHER HALLOWELL. *Green Perspectives: Thinking and Writing about Nature and the Environment*. New York: HarperCollins, 1994.

MORGAN, SARAH, AND DENNIS OKERSTROM. *The Endangered Earth: Readings for Writers*. Boston: Allyn and Bacon, 1992.

ROSS, CAROLYN. *Writing Nature: An Ecological Reader for Writers.* New York: St. Martin's, 1995.

SLOVIC, SCOTT H., AND TERRELL F. DIXON. *Being in the World: An Environmental Reader for Writers.* New York: Macmillan, 1993.

VALENTI, PETER. *Reading the Landscape: Writing a World.* New York: Harcourt, 1996.

VERBURG, CAROL J. *The Environmental Predicament: Four Issues in Critical Analysis.* Boston: Bedford St. Martin's, 1995.

WALKER, MELISSA. *Reading the Environment.* New York: Norton, 1994.

Dropping the Subject
Reflections on the Motives for an Ecological Criticism
ERIC TODD SMITH

In 1989, Cheryll Burgess Glotfelty posed the provocative question, "How can we, as literary critics, respond to the environmental crisis?" In her early and influential paper, "Toward an Ecological Literary Criticism," Glotfelty notes that "[w]hile other social movements, like the civil rights and women's liberation movements of the sixties and seventies, have had a significant impact in shaping literary studies, the environmental movement of the same era has not" (2). To fill this gap, Glotfelty seeks to "mobilize the literary community on behalf of the larger biotic community" (2) with an "ecological criticism, or ecocriticism" which would study "the *relationship* between human culture and the environment" (5). In Glotfelty's argument, and in many of those who have followed her call for an "ecocriticism," the *"relationship* between human culture and the environment" is sick, misguided, and therefore at a turning point or crisis.

This is an important statement and one that reflects a familiar ontological universe. Specifically, by identifying ecocriticism as the study of "the *relationship* between human culture and the environment," Glotfelty's definition makes real the distinction between "culture" and "nature," granting each term status as a coherent noun. The idea of a primary separation between culture and nature has analogues in Western history and epistemology: self/other, subject/object, mind/body. Ecocriticism responds in part to the sense that such dualisms result in abuse—the subject disdains mere objects, culture objectifies nature. Indeed, Glotfelty, like many other ecocritics, is keenly aware of the hazards of dualistic thinking. As she notes in her recent introduction to *The Ecocriticism Reader*, "Some scholars. . . . favor *eco-* over *enviro-*" as a prefix to their brand of criticism, because "*enviro-* is anthropocentric and dualistic, implying that we humans are at the center, surrounded by everything that is not us, the environment. *Eco-*, in contrast, implies interdependent communities, integrated systems, and strong connections among constituent parts" (xx). A primary motive for ecocriticism, then, has been to find a way out of anthropocentric and dualistic thinking. The fact that Glotfelty's early definition of ecocriticism reproduces the dyad of "human culture and the environment" underscores the seriousness of the challenge before ecocriticism; the assumption of a dualistic universe—a universe divided between subjects and objects—resides in the very terms we use to talk about "the environmental crisis."

By comparing ecocriticism to "other social movements," Glotfelty asserts a principle of enfranchised subjectivity as both a goal and a measure of fairness ("Toward" 2). Indeed, the civil rights and women's liberation movements have been successful to the extent that they have demanded recognition of subject status for people who were once recognized only as objects. Subject status in these "emancipatory movements" has been synonymous with having a voice, "having a say" in things. Following this logic, a number of ecocritics have sought to rehabilitate the "voice of nature" and to listen carefully to what it tells us. "Nature," in other words, is brought into the category of "subject" with the hope that its "interests" will become recognizable.[1] This motive, the desire to work against the objectification of nature by defining it as a subject, is what I wish to examine in the rest of this essay.

The strategy of positing nature as a "speaking subject" is one way of responding to a common holistic hope for "reconnection" or "unity" between humans and nature. While I will admit to certain holistic fantasies of "connection with nature," I believe that the effort to unify being under the heading of "speaking subject" is something like putting a square peg into a round hole. It seems to me that the first misstep lies in the familiar vocabulary most of us use to talk about this whole issue: by granting dualities like culture and nature or subject and object, we also grant a basic ontological disparity, a polarity with nothing in between. My intent in this article is to imagine that we might step out of a universe constantly straining between two grand epistemological poles of subject/object, or culture/nature, and instead stand in the midst of a universe of relationships between entities that constantly mediate and translate each other. This is not an attempt at revolution, as the word "dropping" in my title implies, so much as a change in the directions and valences of critical attention. Purity here is meaningless; we shall never find the clearly drawn boundary between culture and nature or subject and object, although we will be able to make distinctions between relationships. Similarly, boundaries between literature and other kinds of mediation become less clear. Glotfelty's intent to study "the *relationship* between human culture and the environment" is an important step in this direction. Even so, I think we first ought to examine the idea of disparity defining the two poles of this relationship. Then we might more easily see how some ecocriticism suggests a different critical position even while effectively precluding it.

The mission for ecocriticism, as Glen Love puts it, is to "revalue nature," significantly by advocating "nature-oriented literature" as a "needed corrective to . . . anthropocentric . . . assumptions and methodologies" (205–6). The "re-" prefix in the title of Love's manifesto, "Revaluing Nature: Toward an Ecological Criticism," posits a human alienation from nature and an antipathy for nature, both of which must be overcome by the critical project of ecocriticism. These two intertwining theoretical themes pervade much ecocriticism: nature has been unjustly dominated and must be liberated, "given voice"; *and* this domination stems from an alienated, pathological "anthropocentric" ideology that must be cured by reconnecting human culture with nature. This indictment of anthropocentrism—the idea that human attention directed only toward human culture results not only in unjustifiable "domination" of "nature," but also in a particular malaise in human society—points to the philosophical holism undergirding many ecocritical statements to date. While the notion of "connection" between subject and object usefully resists dualistic thinking, the problem is that this union is presented in terms of nostalgia: the subject can only choose between current alienation and restored unity. A basic ontological disparity between subject and object, in other words, is taken to be factual and inevitable. In this respect, nostalgia for a lost unity with nature reflects the attitude best described as modernity—the sense that our world is utterly different from the world of the past, the sense that we have been alienated from the "state of nature."

Whether or not a "lost unity" between culture and nature is taken to be an epistemological given—many postmodern critiques, for instance, would never make an assertion so resonant with "presence"—the poles of culture/nature or subject/object persist as a structure for conceptualizing the "environmental crisis": culture is doing something bad to nature, the self is oppressing the other with "instrumental reason." In other words, objects are abused because they are not valuable to subjects. A similar line of reasoning has been useful to feminists critiquing pornography. Such a model casts the crisis in terms of an imbalance of power and suggests that doing away with objectification will equalize power, thereby "liberating" former "objects." Many ecocritics seek to do this by recovering the subjecthood of nature—that is, by moving all of nature from the realm of the object into the ideal sphere of the subject. This view assumes that all significant power lies within subjectivity, and it *is* true that the social movements Glotfelty refers to have had relative success because they have demanded "subject status" for a previously invisible class of "objects." Women, ethnic minorities, gays, and lesbians, for example, have all increasingly gained a "voice" as legitimate subjects in society. Even so, the insistence on subjectivity as the measure of validity becomes problematic when the potential "subjects" in question—ecosystems, migrating waterfowl, the Earth—do not share human language. Conflicting accounts of what nature is telling us lead quickly to an argument about who *really* understands what nature wants, an argument impossible to resolve. It seems to me that insisting on subjectivity as the route to validity is an example of what Donna Haraway calls the "reductionism" in ideas of "universality . . . when one language . . . must be enforced as the standard between all translations and conversions" (187). The challenge, then, is to imagine relationships between humans and other entities outside the standard categories of subject and object or culture and nature. By focusing on relationships, even if they are conceived between the poles of subject and object, ecocritics have begun to reimagine unity.

As long as the necessary ecocritical move is seen to be the emancipation of a nature's smothered subjecthood, critics are forced to theorize ways to accommodate "nature" (suddenly unified into a coherent, if silenced, subject) in the arena of human political language. In order to avoid the bugaboo of "anthropocentrism" with this move, ecocritics must posit different definitions of language and subjectivity that ostensibly extend beyond human expression. Patrick

Murphy has responded to this problem with a "feminist dialogics" derived from Bakhtin, which he suggests "requires a rethinking of the concepts of 'other' and 'otherness' [with] the corollary notion of 'anotherness,' being another for others" (149). By adding this reciprocal construct to otherness, "the ecological processes of interanimation—the ways in which humans and other entities develop, change, and learn through mutually influencing each other"—become visible (149). The concept of "anotherness" is related to Bakhtin's conception of "the individual as a *chronotopic relationship*, i.e., a social/self construct within given social, economic, political, historical, and environmental parameters of space and time" (150), which means that "[t]he 'other' in its various manifestations . . . participates in the formation of the self" (151). The concept of the individual as "chronotopic relationship" approaches the connection ecocriticism seeks between self and other, culture and nature, by acknowledging the embeddedness of any subject in a context of lived experience and thereby attempting to dissolve the boundaries between subject and object. The notion of the other and the self participating in each other's formation is valued by Murphy in somewhat the same way that "nature-oriented literature" is valued by Love—as a "needed corrective" (Love 205), a way of bringing the object (nature) into the subject sphere (the human sphere).

Yet, "participation" of the other in the formation of the self becomes problematic when extended to subject-based language:

> Just as the "other" participates in the formation of the self, so too does the self as individual-in-the-world participate in the formation of the "other" in its various manifestations. And just as that self enters into language . . . so too does the "other" enter into language and have the potential, as does any entity, to become a "speaking subject." [The "other" is] constituted by a speaker/author who is not the speaking subject but a renderer of the "other" as speaking subject. (Murphy 151)

In order to figure the "other" as a speaking subject, Murphy must posit a "channeler" of sorts, who will deliver the subjecthood of the other into language. "The point," Murphy continues, "is not to speak for nature but to work to render the signification presented us by nature into a verbal depiction by means of speaking subjects, whether this is through characterization in the arts or through discursive prose" (152). The difference between "speaking for" nature and "rendering" nature is distinguished here by a certain faithfulness of transmission, a desire not to *misrepresent* the "other." At work in this effort to listen to the previously silenced voice of nature is the same ideal of enfranchised subjectivity motivating the "emancipatory" movements upon which Glotfelty models ecocriticism: if nature's voice can be heard, then it can "have a say" in things.

While Murphy explicitly claims that nature need not have volition, as humans do, to speak, he is concerned to preserve an autonomous authenticity of nature, asking, "When selenium poisons groundwater, causes animal deformities, and reduces the ability of California farmers to continue to cultivate, are these signs we can read? And in reading such signs and integrating them into our own texts, are we letting the land speak through us or are we only speaking for it?" (153). I agree with Murphy's intent, which is to make the consideration of environmental dangers, and the power structures that lead to them, a critical moral issue and not just another idea lost in pluralist relativism. But I resist Murphy's effort to grant speaking-subject status to nature because it rests on the idea that speaking is the best way to gain legitimacy. By basing the "constitution of nature as a speaking subject" on a "rendering" proxy, Murphy hopes to establish the considerability of nature as a subject whom humans are oppressing with pollution, urbanization, population, etc. I certainly agree that humans can interpret and understand the signs of pollution and environmental degradation, and that such interpretation is determined by emotions, moral values, and sympathy for the suffering of animals. But when the authority for assessing the "environmental crisis" is centered in a pure, if silenced, subject—nature—with whom we must communicate, the discussion has simply been deflected toward a debate over which "proxy" legitimately represents nature's interests. My skepticism does not dispute the moral considerability of nonhuman entities, but rather suggests that subjectivity as a model for determining value, moral considerability, and action is limited. In the end, the question of "what the land means" carries only as much weight as the person arguing for it; conflicting accounts of what nature "means" will persist and the ensuing arguments will necessarily be over who has perceived the authentic meaning.[2]

Peter Quigley has addressed this issue of authenticity directly, suggesting that it is difficult to include "nature" in current "rational" conceptions of language and subjectivity without converting it into "[t]he illusion of a free and unencumbered individual," the speaking subject of liberal humanism (299). The problem is that no matter how many oppressed subjects can be "emancipated" from the object sphere of nature, the polarity of subjects and objects still leaves only two primary ontological categories with a vast gap in between. In this view, "subject" and "object" or "society" and "nature" seem like remarkably homogenous classifications for the amazing variety of entities and relationships in the universe.

Perhaps, then, subjectivity should not be the goal. I suggest we drop the subject of the subject, and that of its defining opposite, the object, as the grand poles staking out existence. Let us think, rather, about multiple mediations and relationships, not marked out by membership in one of the two great camps of subject and object, but rather by specific embodiments, situations, and affinities. Instead of seeing "nature-oriented literature" and criticism as ways of giving "voice" to nature, I propose that ecocritics think about literature and criticism as simply particular kinds of relationships between things (and people *are* things, like everything else). By suggesting this ontological perspective I hope to avoid the infinite regress of competing claims about what nature-as-a-subject really wants. At the same time, I hope to find a way to talk seriously about the way people value relationships with nonhuman entities, even if we can't prove definitively what they are telling us. I don't think, in other words, that being a subject in human language ought to be the model category of ethical considerability.

The difficulty of imagining relationships without the clear categories of subject and object is indicated by what Bruno Latour calls modernity's "work of purification." Latour sees the central labor of modern society as the labeling and organizing of existence according to how things fit the two poles of subject and object or culture and nature. In the modern view, claims Latour, entities that do not fit into one or the other of these two categories are seen as "hybrids [and] monsters" (47), that is, *"as a mixture of two pure forms"* (78), suspended in ontological limbo between the purified extremes of subject and object. Non-"purified" entities, in other words, can only be defined by their failure to be either entirely natural or entirely cultural. "In the modern perspective" (80), claims Latour, the only accepted relationship between subjects and objects is through "intermediaries," such as technologies of observation or measurement, which allow the subject to remain independent—clearly distinct—while observing nature, the object. Such intermediaries "establish links [between subject and object] only because they themselves lack any ontological status" (80).[3] Within this framework, which Latour calls the "modern Constitution," there is no legitimate existence outside the categories of subject and object.

Latour suggests that we look closely at the no-man's land between the purified poles of subject and object. In his explanatory model, which he terms "nonmodern" (as opposed to modern or postmodern), "[t]he point of separation—and conjunction—[between subject and object or culture and nature] becomes the point of departure. The explanations no longer proceed from pure forms toward phenomena, but from the centre toward the extremes" (78). Instead of imagining intermediaries, which keep the two poles distinct, Latour proposes that we imagine "mediators": "[A] mediator . . . is an original event and creates what it translates as well as the entities between which it plays a mediating role" (78). A mediator is not subordinate to other, more ontologically valid entities, but rather is an entity, like all others, existing in and through relationships. Once mediation is taken to be not only the nature of relationships between entities, but also the *constitutive activity* of entities, "we notice that there is no longer any reason to limit the ontological varieties that matter to two" (79), such as culture and nature or subject and object. Because mediation is the nature of existence, purity is not valued, much less possible. Nothing can be entirely of nature or entirely of culture. Every *thing* is a mediation and translation of other things, a relationship not void of being but constituted by relations.

Even if pure states of being are no longer available, we can, of course, continue to make distinctions and value judgments. We no longer need to determine whether nature has been unduly altered by culture, or whether culture needs the influence of nature. Rather, we can ask what kinds of mediation are going on, how we might change them, and how they might contribute to what Donna Haraway calls "projects of finite freedom, adequate, material abundance, modest meaning in suffering, and limited happiness" (187). A literary critic, then, can be seen as an interpreter of and participant in complex relationships who speaks from his or her embeddedness in those relationships, not from an imaginary "critical distance." From this perspective, Murphy's question—about whether the land is "speaking through us" when we write or talk about ground-water pollution and animal deformities—appears slightly misdirected. The point is not finally to determine whether we can read nature's will, but to speak as participants in what Glotfelty calls our "immensely complex global system, in which energy, matter, *and ideas* interact" ("Introduction" xix). When we read the signs of environmental crisis and "integrat[e] them into our own texts" (Murphy 153), I do not think we are rendering the authentic subject status of a wounded Earth. Instead, I think we are expressing and enacting our relationships with land, pollution, animals, anger, and desire, among other things. As I see it, these relationships are the basis of value, and, I hope, political action.

Donna Haraway's image of the "cyborg . . . a hybrid creature, composed of organism and machine" (1), illustrates the power of defining value not according to subject status but rather on the basis of ongoing relationships. Because it exists entirely in the wasteland between the modern categories of culture and nature, subject and object, yet persistently "refuse[s] to disappear on cue" (177), the cyborg forces an alteration of dualistic schemes of existence: "Nature and Culture are reworked; the one can no longer be the resource for appropriation or incorporation by the other" (151). The image of the cyborg can be extended to reflect the inadequacy of the polarized, purifying "modern Constitution" Latour describes. Haraway explains, "[c]yborg imagery can suggest a way out of the maze of dualisms in which we have explained our bodies and our tools to ourselves. This is not a dream of a common language, but of a powerful infidel heteroglossia" (181). By stressing that entities need not share a "common language" in order to share valid and valuable relationships, Haraway points up the false choice between total subjectivity and total objectification implied in attempts to grant speaking-subject status to nature. The lack of a "common language," however, does not preclude the existence of "an earth-wide network of connections, including the ability partially to translate knowledges among very different—and power-differentiated—communities" (187). Such "partial translation" among different entities is as imperfect as our language, but it casts a wider net and brings more entities into what Latour calls the "Parliament of Things" (142), the collectivity of all beings. There is a kind of unity in this view; entities *are* inevitably linked together by networks of mediations and relationships that are undeniably real and not merely "discourse effects." But it is not a unity of nostalgia, for there has never been a separation between subject and object.

Glotfelty's call to study the *"relationship* between human culture and the environment," then, is an important first step toward a condition we have never really left. We should study relationships, yes, but

I hope we might begin to do so without first sorting existence into the categories of subject and object. If we understand ourselves as mediators as well as mediated, then we don't have to worry about "rendering" a pure but silenced other, because we will understand that there are more relationships than just linguistic ones. We won't consider our language to have failed when it doesn't deliver the essence of a "referent," because we will know that language is itself a constitutive relationship. This means that literary critics can "respond to the environmental crisis" ("Toward" 1) by refusing to preserve literature as a pure salve (either natural or metaphysical) for the alienated human soul. Rather, ecological critics are in the position to insist upon the way literature (like so many things) extends beyond any arbitrary boundary of art or academe through a network of vivifying relationships. "Pure" art may evaporate along with the polarity between subject and object, leaving not a void, but the infinite variety and possibility of our shared existence.

Notes

1. This philosophy gains perhaps its fullest articulation in Roderick Nash's *The Rights of Nature*. Nash argues that figuring nature as having "moral standing" (7) is part of a "historical tradition of extending rights to oppressed minorities in Britain and then in the United States" (6).
2. With more time and space, I could cite other examples of ecocritics attempting to give voice to nature, notably from Christopher Manes and David Abram, but I think the point is clear. I want to stress again that I do not dispute the underlying ethical concerns of these critics. I suggest, rather, that because it is difficult to agree on what nature-as-a-subject is saying, the ideal of subjectivity is an unreliable basis for effective ecocritical praxis. This all comes down to a familiar problem of representation and authenticity: How do we know if nature is being rendered "correctly"? I hope to suggest that this question needlessly simplifies our relationships—biological, moral, and aesthetic—with other entities.
3. The example to which Latour repeatedly returns is Robert Boyle's vacuum pump, which has sometimes been cast in the history of science as a mere tool which allowed the disovery of certain "realities" concerning the properties of gasses. In this perspective, Latour claims, the "tool" undergoes a curious ontological subordination to

the ostensibly purer "reality" it points to; the pump has been understood as simply an "intermediary" between the scientist and the realties of nature.

Works Cited

ABRAM, DAVID. "Merleau-Ponty and the Voice of the Earth." *Environmental Ethics* 10 (Summer 1988): 101–20.

GLOTFELTY, CHERYLL. "Introduction." *The Ecocriticism Reader.* Ed. Cheryll Glotfelty and Harold Fromm. Athens: U of Georgia P, 1996. xv–xxxiii.

———."Toward An Ecological Literary Criticism." Western American Literature Meeting, Coeur d' Alene, Idaho, 13 Oct. 1989.

HARAWAY, DONNA. *Simians, Cyborgs, and Women: The Reinvention of Nature.* New York: Routledge, 1991.

LOVE, GLEN A. "Revaluing Nature: Toward an Ecological Criticism." *Western American Literature* 25: 6 (1990): 201–15.

LATOUR, BRUNO. *We Have Never Been Modern.* Trans. Catherine Porter. Cambridge, MA: Harvard UP, 1993.

MANES, CHRISTOPHER. "Nature and Silence." *Environmental Ethics* 14 (Winter 1992): 339–50.

MURPHY, PATRICK. "Ground, Pivot, Motion: Ecofeminist Theory, Dialogics and Literary Practice." *Hypatia* 6.1 (1991): 146–61.

NASH, RODERICK. *The Rights of Nature: A History of Environmental Ethics.* Madison: U of Wisconsin P, 1989.

QUIGLEY, PETER. "Rethinking Resistance: Environmentalism, Literature, and Postructural Theory." *Environmental Ethics* 14 (Winter 1992): 291–306.

Bodega Head
An Excursion in Nuclear Shamanism

JOHN P. O'GRADY

They called it the "Glory Hole." Seventy-three feet deep, one hundred forty-two feet in diameter. It took two expensive years of excavation to open the hole, the work being completed in October 1963. Spoils from the site had been used to construct an access road across the tidal mud flats at the northern end of the bay. Along the scarified shoreline of Campbell Cove, heavy equipment stood poised, ready to begin the next phase of construction. Final federal approbation, expected at any time, was all that remained between this and the earnest work.

The place, about sixty miles north of San Francisco, is called Bodega Head, but on a map it looks more like a thumb. From a wide, undulating east-west trending series of sand dunes that connects it to the mainland, Bodega Head is a rocky spit of land, two miles long by half a mile wide, jutting southward into the Pacific Ocean, yet more than sand separates this block of quartz diorite from the sedimentary bulk of northern California. The sand dunes, in fact, make manifest to the discerning eye the rift zone of the San Andreas fault. Geologists speak of California as being "tectonically active," but in the early 1960s this part of the state was active in other ways. There were designs on the Head. It had been bored. It was ready to receive its crown: a $61-million "atomic power plant." With a generation capacity of 325 megawatts, it would have become the largest nuclear generating station in the United States, capable of producing enough electrical power "to serve a city of 500,000 persons" (*San Francisco Chronicle*, 29 June 1961). It was to have been the first nuclear power plant to cross the threshold of commercial profitability.

All of this, however, was not to be.

There was strong grass-roots opposition to the Pacific Gas and Electric (PG&E) plans for a reactor at Bodega Head. Historians have provided us with detailed accounts of the various political machinations, but their stories come to a close at the end of October 1964, when the Atomic Energy Commission (AEC) issued a contradictory pair of reports on the proposed reactor. On the one hand, the Advisory Safety Committee on Reactor Safeguards concluded that the plant could be built "without undue hazard to the health and safety of the public." On the other hand, the Division of Reactor Licensing concluded "Bodega Head is not a suitable location for the proposed nuclear power plant at the present state of our knowledge." This branch of the AEC was responding to the grave concerns some geologists had expressed about the plant's being sited within a quarter mile of the San Andreas Fault. Although this section of the fault is not particularly active, during the 1906 quake that devastated San Francisco lateral ground movements in the Bodega area achieved displacements of up to ten feet. Despite assurances by PG&E engineers that their design would take all this tectonic capriciousness into account, a rogue geologist in the United States Geological Survey had in the previous year written in a report that "Acceptance of the Bodega Head as a safe reactor site will establish a precedent that will make it exceedingly difficult to reject any proposed future site on the grounds of extreme earthquake risk" (*San Francisco News Call Bulletin*, 4 Oct. 1963). AEC final approval was still pending—and may have been imminent—but on October 30, 1964, PG&E withdrew its application for a license to build and operate a nuclear generating station at Bodega Head. "We would be the last," a press release assured, "to desire to build a plant with any substantial doubt existing as to public safety."

Today the "Glory Hole" is a freshwater pond. The scarifications have been recolonized by California coastal prairie and scrub, a fragrant mosaic of grassland, coyote bush, cow parsnip, and poison oak. The edges of the pond are thick with cattails and rushes. Only a keen landscape sensibility would be able to detect signs of the previous and near-plutonic goings-on. The lands that had been acquired by

PG&E for the power plant are now a state park. To see this place today and compare it to the photos from more than thirty years ago gives great encouragement to those who place their faith in "ecological restoration." Bodega Head is an environmental "success story." A beautiful place has been saved; indeed, it has been returned to a condition far wilder than it was in the early sixties when PG&E dug its hole, for in the long decades prior to that the Head had been heavily grazed by horses, cattle, and sheep, the effect of which was to suppress or eliminate most of the native vegetation, especially the annual wildflowers and lupine shrubs. Apparently, all of this has come back: Persephone is once again at play in the daylight amid the blooms.

At present an aficionado of wildness, such as myself, might walk through this recovered patch of pristine California and meditate freely upon the words of poets. Take a few lines from Philip Whalen, for instance:

> . . . here are
> no fields where food is growing, no smell of night-soil,
> here's all this free and open country, a real luxury that
> we can afford this emptiness and the color of dawn
> radiating right out of the ground.

As can be gleaned from its title—"America inside & outside Bill Brown's House in Bolinas"—Whalen's poem is about America. Its narrator is sensitive to the glories of a particular landscape (one that is not many miles south of Bodega Head), but he is also acutely aware of the "luxury" such wild lands afford: the food is grown somewhere else. In the case of Bodega Head State Park, we now can indulge our appetite for wildness there because our electricity is generated on some other piece of land. The economist would call this a "trade-off." The pre-Socratic Greek philosopher Heraclitus speaks to this binary nature of reality with cryptic panache: "The beginning and the end are shared in the circumference of a circle." And Ralph Waldo Emerson, in his essay "Compensation," expresses this fundamental paradox of existence in terms most relevant to the case at hand: "To empty here, you must condense there." PG&E, not ignorant of the Great Western Tradition in philosophy, reminded its customers of fateful inevitability when, in language somewhat less polished than Emerson's, they withdrew their application for the power plant at Bodega Head: "We have made provision for adequate electric generating capacity elsewhere to take care of our customers' needs for the several years immediately ahead. Our decision to withdraw the Bodega application does not mean we have lost any confidence whatsoever in nuclear-electric generation." Indeed, they stood by their words, persevering for more than a decade against strong opposition before finally opening their Diablo Canyon nuclear plant, not far south of California's renowned Big Sur, in a landscape equally spectacular—and tectonically active—as Bodega Head. The Glory Hole on the northern California coast has been filled with water and birdsong; two new glory holes subsequently opened on the central coast have been filled with ingenious machines that split atoms.

I myself am not much interested in "environmental guilt." I'm happy that Bodega Head is now a state park and I never regretted visiting it by car. I'm pleased to drink the milk that comes from the cows grazed somewhere other than here, and I love my sweaters made from wool sheared off "hooved locusts" that denuded some faraway grassy hillside never to enter my ken. Although the poetaster in me is disquieted by the thought of the Diablo Canyon plant, I used with gratitude the electricity that PG&E supplied me when I lived in California, and I always paid my bills on time. The company would occasionally print a message at the bottom of my monthly statement: "We appreciate the opportunity to serve you. Your payment history establishes you as a good credit customer. Thank you." I take this to mean we have good relations, myself and this corporation.

I am not much interested in hedonism either. While acknowledging my own complicity in our postindustrial consumer culture, I do not in this case wish to follow the often misconstrued advice of Martin Luther, "Be a sinner and sin bravely." Nor do I think it a wise retort to invoke Walt Whitman's famous accosting of his reader: "Do I contradict myself? / Very well then I contradict myself." Such rhetorical strategies are coy posturing and ultimately self-erosive, for if both Artemis and Reddy Kilowatt are among the pantheon of gods that cavort in the hearts of late twentieth-century Americans, you can't very well build an altar only to the one and not expect to incur the wrath of the other.

I'm not even interested in resorting to a fashionable "middle ground," because to employ this design is to cleave one dualism into two, each of which will then require a middle ground, which in turn creates two more dualisms, ad infinitum—disturbingly reminiscent of that pattern found in the ungoverned division of cells in an organism, a phenomenon we refer to as "cancer." Nuclear fission, as well, lends itself here as a metaphor. In our use of figures it is easy to get carried away. Consider the boilerplate of scholarly discourse. After casting a discerning eye upon the controversial idea of wilderness and finding it wanting, environmental historians and some literary scholars have recently been issuing enthusiastic calls for everyone to hurry to the commonsensical middle ground. The trouble these thinkers have with wilderness is that it is a "cultural construction" created by Americans so they can flee from responsibility. This makes for a fine scholarly bedtime story, the moral of which is (in the words of one influential professor) that we must all strive "for critical self-consciousness in all of our actions" and never imagine "that we can flee into a mythical wilderness to escape history and the obligation to take responsibility for our own actions that history inescapably entails." Noble words, yet I for one become uneasy when I see a Procrustean bed being made. Historians are wont to insist that theirs is the one academic discipline from which there is no escape. Ironically, this same historian quotes the poet Gary Snyder to make the point that "wilderness" is a quality of one's own consciousness. No doubt, but this being the case, one could just as easily say anybody's notion of "history" is also a quality of one's consciousness. The problem is that any middle ground rising from intellectual plates grinding against each other in the scholar's mind is as unstable as any other in the landscape of argument. Even Bodega Head would provide more reliable ground on which to stand.

Scholarship—as well as the "problem-solving" mindset that is our culture's inheritance from that nineteenth-century intellectual excrescence known as positivism—is in many ways like a water strider, always skimming along the surface. Many of us feel a need to get beneath the surface, to reach, if only with our feet, some new, heretofore concealed place that might serve as foundation for a new consciousness. Yet how does one get beneath the surface? Is it possible to "think" one's way there? Perhaps, but poetry—or more specifically, a poetic style of consciousness—shows there are other ways. In the passage I quoted from Whalen's poem, the narrator refers to "the color of dawn radiating right out of the ground." More than dawn radiates out of the ground in any given place. One does not "think up" stuff such as this, one encounters it—it happens *to* one. Mary Austin frequently wrote about what she called the "exhalations" of the land—certain energies "radiating," if you will, from a particular place, expressive of the life that is lived therein. *These* are the "spirits of place." They are always there but give a nod only to the person who is open to them. In his poem "The Snow Man," Wallace Stevens calls this openness "a mind of winter."

If you desire a fully elaborated "theory" about all this, study the laws of karma; or better yet, meditate on them. Try this at home. Little mention is ever made of meditation in contemporary critical discussions; this comes as no surprise, for meditation would suggest a style of consciousness that, in effect, temporarily suspends critical judgments and requires instead a cold and rigorous plunge into wakefulness, a "shiverednesse of soule all to pieces," as Thomas Hooker, the seventeenth-century Puritan theologian and founder of Hartford, Connecticut, put it. Some of us dive and some of us are pushed, but to plumb the depths of your own soul is not without risk—especially within the academy, where many intellectuals are of the opinion that no such thing exists—yet this risk need not be ventured with dead seriousness either. If you start talking or writing about these things, you might experience a loss of professional respectability but your intellectual life might suddenly become joyous again. No, a water strider will not go beneath the surface, but a loon can dive deep—and come up laughing.

A few years ago I attended a session on "California Power Places" at the Shasta Nation Bioregional Gathering in the hills above the Napa Valley. The woman who led the discussion, Lynne, was a horticulturist by avocation, one who specialized in California native plants; she made her "real money" as a technical writer for one of the computer companies in the Silicon Valley. She was ordinary looking, slightly overweight, in her mid-fifties. She dressed like an old hippie. One got the impression that this was her weekend wardrobe.

To encourage all of us in the group to talk about our own experiences, no matter how strange they might be, Lynne shared a story about her encounter with one of those outcrops of chert (locally referred to as "knockers") on the western slope of Mt. Tamalpais. She told us how when she was walking along through this steep grassland with her friend Bill she suddenly heard what sounded like voices rising from between the rocks. She wanted to stop and investigate. Bill, though hearing nothing himself, was used to his companion's endearing eccentricities, so he provided support.

"I'm going down," she announced. Without further ado she got on hands and knees and—like a dabbling duck dipping in the water—stuck her head down between the rocks, her bottom up toward the sky. The next thing Bill knew, Lynne was flying up and backwards, as if tossed. She landed on her butt in the grass. Fine otherwise, she had a slight bruise on her forehead, what looked like a small dent.

"They whacked me!"

"Who whacked you?"

"The spirits in those rocks."

Lynne reported that when she stuck her head between those rocks, she suddenly found herself looking down, as if from the ceiling of a vast cavern, into a realm where shadowy, gnome-like figures were flitting about in a great commotion.

"They were not happy to see me," she explained. "One of them stood right underneath me and threw a fist up at my forehead. That sent me right back up into our world."

Blessed with a healthy curiosity, Bill himself investigated the rocks but found no portal to the underworld, only the solid earth, yet he believed Lynne's story. Over the years he had seen her tossed up from other rocks and stumps, and one time from an apartment complex dumpster. He had witnessed enough of this ordinary woman's extraordinary encounters with the world to have faith in her experience.

Yes, Lynne, who had a normal nine-to-five job and grew flowers on the weekend, was "crazy"—I came away from the session with a profound sense of *that*—but it was a refreshing craziness, one that punctuates day-to-day routine with a sense of wonder. Something blossoms in Lynne that is missing from the academic, corporate, and government worlds that comprise our "mainstream" culture—namely, a playful style of being that does not lay waste to the full human power of perception. She regularly obtains glimpses of things that leave her open to public ridicule and secret envy. She is nuts. So I told her—why not?—about the time I saw a whole New England graveyard of ghosts leap up from their plots and chase me as I drove my car down an isolated road. I was terrified, not so much of the ghosts but that somehow it would be found out that I saw ghosts. I drove on toward my destination with hasty determination.

"They must have had something important to tell you," Lynne now said to me with great conviction. She chastised me for not stopping for those spirits radiating out of the eastern Maine landscape who, for reasons that remain tucked in their graves, wanted to have a word with me.

I laughed, a bit nervously, at Lynne's ardor, but I came away from that session energized in a way I have never felt at any academic conference. Perhaps all this served as a catharsis. In the end, can it be said that this woman is any more crazy than those who would build nuclear power plants along active earthquake faults, or those who would "reinvent nature" in the windowless, air-conditioned seminar rooms of our universities?

The young Mary Austin in her journal wrote about "a sensual way of beholding." When it comes to the land, she insisted that "There is something else there besides what you find in the books; a lurking, evasive Something, wistful, cruel, ardent; something that rustled and ran, that hung half-remotely, insistent on being noticed, fled from pursuit, and when you turned from it, leaped suddenly and fastened on your vitals. This is no mere figure of speech, but the true movement of experience." John Muir occasionally mentions angels in his writing; there is evidence to suggest he was speaking literally. Are we to say then that Austin and Muir, like Lynne, are a little wacky? Yes, yes, yes—decidedly more so than the psychologically sanitized commentaries upon their work that we are now producing.

Our everyday lives and especially our literature are full of kooks, no question about it, necessary kooks. The kook, in his or her kookiness, is a fringe figure, outside any ideology, in some way akin to the shamans of old, whom Mircea Eliade in his classic study, *Shamanism: Archaic Techniques of Ecstasy*, identifies as the precursors to what we today call poets. All of these crackpots—shamans, poets, pilgrims to California Power Places, and even the nuclear engineers who dream their Promethean dreams—direct our attention to what Eliade describes as "the fabulous world of the gods and magicians, the world in which *everything seems possible*, where the dead return to life and the living die only to live again, where one can disappear and reappear instantaneously, where the 'laws of nature' are abolished, and a certain superhuman 'freedom' is exemplified and made dazzlingly *present*." These full-souled people living on the edge are neither reasonable nor timid; they are, after all, "enthusiasts" in the root sense of the word, that is, "filled with a god." In reading our eminently reasonable environmental histories and "ecocriticisms," the soul in its quest for meaning encounters a debilitating resistance to its progress. No wonder it resorts elsewhere for inspiration. To places like Bodega Head, or Yosemite, or Yellowstone, or any of our state-sanctioned "wilderness areas." Yet, can we say with assurance that even these places serve?

Because it is now "wild" and has a luminous environmental history, Bodega Head serves as a spiritual *locus minoris resistentiae*, a "place of lesser resistance," where all the gods and demons, who drive each of us around like a Winnebago, can get out and stretch their legs, leaving the "vehicle"—that is, the ego—parked and emptied. Since you yourself are the vehicle, you ought to be alert to when this "emptying out of consciousness" occurs, because it is then that you will see just what a heap of trouble you are, this "person" who is amalgamated of so many predilections, ideologies, and other moods that if you tried to inventory them you would only get lost for trying. If you don't like the vehicle metaphor, abandon it and call yourself a "spiritual questor," or a "pilgrim to the wild." Unfortunately, no matter which analogy you finally apply, these "wild" places, the objects of your affections, each a spiritual *locus minoris resistentiae*, are themselves inextricably caught in the same sticky web of dualism from which so much "environmental" literature, philosophy, and criticism seeks to extricate us.

Thus I was too hard earlier on those earnest thinkers who, with intellectual wedge and mallet, continually cleave one dualism into two. Such "fracturing" is simply indicative of the psyche itself, which—despite our academic bedtime stories known as psychology—we in our everyday lives continue to imagine as being "whole" and uncarved but by experience know to be exposed to a relentless weathering that reduces it, at last, to a jagged heap of stones, talus at the foot of a spectacular cliff. As professional thinkers, we sign our names to a wide range of cockamamie testimonies, everything from the death of the author to the cultural "reconstruction" of that author. Scholars today pose as the plastic surgeons of the mind.

I've been hammering away myself, with my own sledge, here in this essay, this errancy, which began as history but has now crossed the unmonitored frontier into foolishness. I am lost in a fascinating world of rhetorical distinctions, shopping, like everybody else in Allen Ginsberg's "Supermarket in California," for "images." Another bedtime story, which serves merely as hors d'oeuvre to the great nightly feast.

The multiplicity of the world is its own compensation. Unaccountable things happen to upset our willful systems—ungovernable occurrences such as dreams, which were so important not only to the shamans that Eliade writes of but also to the ancient Greeks and Romans, who themselves were still close to shamanistic cultures. Dreams, after all, are the stuff from which those pied pipers Marx and Freud fashioned their respective lifeworks.

Of course I had a dream after visiting Bodega Head. It goes like this:

> *I am standing on the edge of the pond that was the "Glory Hole," its waters sparkling with sunlight. Redwing blackbirds call out to each other from atop the cattails and rushes. Suddenly a giant human-like figure begins to rise, headfirst, from the water. The head is monstrously large and out of proportion to the body. Atop the water the figure now stands, covered with mud. Swallows are stitching innumerable circles in the air around the figure's very big head. Suddenly the swallows transmogrify into dark and fast-moving cumulus nimbus clouds; they start gathering furiously. Sparks begin to issue in the air around the figure's head. The whole world darkens and heavy rain falls. The mud is washed away, revealing the figure to be none other than Reddy Kilowatt, that trademark of the power industry. He is smiling, or is it laughing? Gone is the Glory Hole, all of this now taking place in a vast electrical switchyard right out of my New Jersey childhood, when my father worked for the power company. Overhead in the dark sky is an infernal grid of transmission lines, on the horizon nothing but transmission towers. A voice says: "It's a risk worth taking, all of this is beautiful." I awaken with tears.*

Ethnographers nearly a century ago reported that, among some of the Indian peoples of Northern California, a man or woman became a shaman only after dreaming at home of the spirits. At that point, the dreamer could call in a doctor—usually an elder relative—in order to obtain his or her "spirit heritage" (i.e., receive instruction for right conduct with these beings), or the dreamer could go at once to the mountains simply to seek and appease the spirits who had come for a visit. I dreamed my dream at home—long after my last visit to Bodega Head. I have no relatives who are shamans; the only doctors I know are M.D.s and Ph.D.s. I suppose I could have gone to the mountains—or back to Bodega Head—to appease the spirit who came to me as Reddy Kilowatt, but I believe the dream was telling me I had to search elsewhere. I won't find Reddy Kilowatt in the places "untrammeled by man." Instead I need to go to Diablo Canyon or to the Salem Nuclear Generating Station in New Jersey or maybe just to the next conference of the Association for the Study of Literature and Environment. Somehow I don't think any of this is quite right either.

"Show me Thy glory!" Moses shouted to his god from the wilderness area atop Mount Sinai. Now other gods are begging our attention. No need to shout. They too have their glory, and they are near at hand.

PART II
GENRE, GENDER, AND THE BODY OF NATURE

"Whole Shoals of Men"
Representations of Women Anglers in Seventeenth-Century British Poetry

ANNE E. MCILHANEY

I

In British literature angling has long been considered a "contemplative recreation," an activity through which the angler withdraws from society, experiences the peacefulness of nature, of quiet, of solitude, and practices an art worthy of his or her time. In the current decade, an increasing interest in women and the environment has led to a new interest in women anglers and their writings—an interest reflected in the publication of numerous anthologies and essays by and about nineteenth- and twentieth-century women anglers.[1] But women anglers have appeared in British literature for centuries. In the seventeenth century, representations of women anglers proliferate in pastoral and georgic poems, yet these depictions have been largely overlooked in current considerations of the woman angler. This essay, then, attempts to revive the image of the seventeenth-century woman angler and to explore the significance of this image within the context of the larger genre to which it belongs, the British piscatory (fishing) pastoral.

Women did, indeed, fish in seventeenth-century England. Lady Margaret Hoby, for example, mentions angling in 1600 in her diary: "I . . . walked a fisshinge with a freind that Came to me for that purposse" (123; see also 121). Celia Fiennes, too, writes in her 1698 book of travels: "I went to a poole in the Kanckwood 3 mile to fish" (148; see also 131, 166). And in her *The Accomplisht Ladys Delight* (1675), a conduct book for "ladies," Hannah Woolley includes a treatise on angling—which is, according to her, "a recreation which many ladies delight in" (n.pag.). Furthermore, various georgic poems of the seventeenth century depict actual women angling on specified lakes and streams. Even Izaak Walton mentions the "minnow" and the "sticklebag" as fish appropriate for the sport of "Young anglers, or boys, or women that love that recreation" (349; see also 351).

But in seventeenth-century poetry written by men, the woman angler is often depicted through a standard progression of images, many of which appear in Edmund Waller's brief lyric poem, "Upon a Lady's Fishing with an Angle." The speaker of this poem describes "fair Clorinda" sitting alongside a stream, angling (line 1). Envious nature—the "painted flowers," "the streams, the meadows"—attempts unsuccessfully to surpass the woman's beauty (4, 8). Clorinda fishes skillfully: "she makes the trembling angle shake"; "With careful eyes she views the dancing float" (11, 23). Yet her skill seems unnecessary, for the fish "Willingly hang themselves upon her hook" (16); the fish are a part of the nature that she so easily commands. From her external appearance, one "would swear she's made of perfect innocence" (30). But the speaker knows better: "she's masked under this fair pretense"—she hides a hook under the "bait" of her loveliness (29). For the speaker, in the end,

> Her beauteous eyes ensnare whole shoals of men,
>   Each golden hair's a fishing line,
>   Able to catch such hearts as mine,
> And he that once views her bewitching eyes,
> To her victorious charms (like me) must ever be a prize. (36–40)

The representation of the woman angling for fish leads effortlessly into the metaphor of the woman as angler of men, and of the speaker himself, who warns others of the "danger." In this emblematic portrait, Waller encapsulates the image of the seventeenth-century woman angler as she finds representation in poetry by men.

The poetic image of the seventeenth-century woman angler belongs to the tradition of the British piscatory pastoral—a popular genre of the time, and one that is from the outset allegorical. The dominant allegory of the British piscatory is religious: the angler, who carefully makes his fishing tools and then fishes in the solitude of nature—an activity conducive to contemplation—becomes a "fisher of men" in the tradition of the Christian apostles.[2] The piscatory of the woman—a less frequent image which is, as in Waller's poem, also largely allegorical—plays on and modifies this dominant mode of the "spiritual fisherman." For the woman angler, too, fishes along lakes and streams, amid the beauty of nature. The woman, too, wields fishing lines and rods and hooks. And finally, she, too, becomes a "fisher of men"—of sorts. But the woman angler transposes the genre from an expression of spiritual fishing to one of erotic fishing. With the woman, the game of fishing becomes, allegorically, the game of love.

II

The seeds of the dominant religious allegory that flourishes in the sixteenth- and seventeenth-century British piscatory are found in the first known printed English angling treatise. The authorship of this work, the *The Treatyse of Fysshynge wyth an Angle* (1496), is traditionally (and probably erroneously) attributed to a woman, Dame Juliana Berners, who was said to be a noblewoman or a nun at Sopwell Abbey in the mid-fifteenth century.[3] In this treatise, the author explores the pleasures of nature that the angler may enjoy—"his holsom walke and mery at his ease, a swete ayre of the swete savoure of the meede floures. . . . He seeth the yonge swannes, heerons, duckes, cotes and many other foules with theyr brodes" (59). At the same time, she gives specific and detailed instructions for making fishing tools and for angling properly. For this author, the proper enjoyment of natural surroundings and the proper methods of angling parallel an ordered inner life. Fishing becomes a means of achieving various types of gain—of becoming "holy, helthy, and sely" (60), as well as of avoiding vices, and engaging in prayer (82).[4]

In the seventeenth-century piscatory, angling is endowed with many of the same characteristics. In his long georgic poem "The Secrets of Angling" (1613), John Dennys, for example, emphasizes the order and beauty of nature, which leads to the contemplation of God. After describing his surroundings in descending order, from the sky downward through the mountains and hills, and eventually to the river, which runs into the sea, he thinks "how strange and wonderful they [the elements of creation] be," such that "whiles he lookes on these with joyfull eye, / His minde is rapt above the starry skye" (book 1, lines 324–35). Izaak Walton's Piscator, too, meditates on the beauty of his natural surroundings (indeed, he quotes these very verses of Dennys), so that from his perspective, "the Rivers side is not only the quietest and fittest place for [Christian] contemplation, but will invite an angler to it" (193). Along with Dennys and Walton, most angling authors of the seventeenth century emphasize an enjoyment of nature as it leads to thoughtful contemplation.

These seventeenth-century piscatory authors give detailed and orderly instructions about how to fish—about the making of fishing tools and the proper methods of angling. At the same time, they give long lists of the "inward qualities," the virtues which the angler should have, to parallel the outward order of nature and of skillful angling. For Gervase Markham, this list includes not only such commonly listed virtues as "faith, hope, and love," but also such accomplishments as knowing the liberal sciences, being a grammarian, having "sweetnesse of speech," knowing the celestial bodies and "Countryes," being "skilfull in Musique," and others (15–17). As the angler's personal qualities reflect the proper ordering of tools and techniques—as well as the order of nature, which leads to the contemplation of God—the angler becomes also, either implicitly or explicitly, a "fisher of men" in the spiritual sense.

III

It is in light of this dominant allegorical piscatory tradition that the less-frequent woman angler passages are most productively examined. For the piscatory of the woman transforms the genre from within; it shares many of the external characteristics of the dominant piscatory, but reworks the allegorical element. This shift, of course, brings changes both to the points of emphasis in the description of the allegorical agents and to the significance with which these agents are invested. For example, whereas in the dominant piscatory the observation of nature leads to the contemplation of God, in the feminine piscatory all of nature adores the beautiful woman. In the dominant piscatory the careful crafting of tools and painstaking skill—along with innumerable inner virtues—are paramount to the success of the fisherman, whereas the woman angler catches fish largely by virtue of her beauty and charm. And while the male angler becomes a spiritual fisher of "men," the woman angler becomes a physical fisher of *men*.

Like the piscatory poems concerning men, the fishing pastorals about women emphasize the pastoral world of nature. But whereas in the dominant piscatory the fisherman observes his natural surroundings in a way that leads him to contemplate God, in the piscatory concerning the woman, all of Nature envies, adores, bows to, or seeks to reproduce the beauty of the woman. Thus, in Waller's "Upon a Lady's Fishing," "The streams, the meadows yield delight, / But nothing fair as her [Clorinda] you can espy" (lines 8–9). In Donne's well-known poem "The Baite," the river is "Warm'd by thy [the woman's] eyes, more then the Sunne" (line 6). Or again, in Stanley's "Sylvia's Park," "Heaven it self conspires / O're all the World to paint her [Sylvia] forth!" In this poem, nature observes the woman's features, which it then paints in its own elements: "In the bright Sun her eyes are drawn; / In the fresh Beauties of the Dawn, / Those of her blushing cheek appear" (lines 83–87).[5] The allegorical shift of the feminine piscatory emerges even in the personification of Nature in these passages. For, in contrast to the Nature of Dennys or Walton, which points ultimately and spiritually to God, the Nature surrounding the woman angler points to her: she enters the poem as an irresistible presence, as an unequaled beauty. Nature's shift in focus—from the Divine to the Woman—parallels various other shifts in the piscatory that accompany the move from male to female angler.

Among these is a shift in the depiction of the actual process of angling. In the dominant piscatory, the angler must construct his fishing implements with great care and precision, and must follow precise procedures as he angles in order to ensure the catching of fish—a process which allegorically suggests ordered living. The woman angler, too, possesses fishing tools and seems to follow the instructions given in various treatises of the day. In Waller's "Upon a Lady's Fishing," for example, Clorinda "makes the trembling angle shake" and "sits as silent as the fish" as she carefully "views the dancing float" (lines 11, 27, 23). Similarly, in Stanley's poem, Sylvia "with one hand the Line she cast" (line 101). And other poems mention the woman's use of the "hooke" (Drayton, line 252) and "line" (Waller, "On St. James," line 35).[6] But in the piscatory of women, the making and use of tools are rarely emphasized; they are rather "props" used to aid in the revelation of the woman's beauty. And it is, in the end, her beauty which ultimately does the work of "catching."

Unlike the male fisher, who must have, along with his skills and tools, a plethora of inward virtues in order to be both a fisher of fish and a "fisher of men," the woman in piscatory poetry needs only her beauty and charm to fulfill both her "real" and her allegorical tasks.

For, like the rest of nature, the fish submit willingly to the woman, choose captivity at her hands over freedom in the water. In Waller's "Upon a Lady's Fishing," "now the armour of their [the fishes'] scales / Nothing against her [Clorinda's] charm prevails," so that they "Willingly hang themselves upon her hook" (lines 14–16). In Donne's "The Baite," the fish "Will amorously to thee swimme, / Gladder to catch thee, then thou him" (lines 11–12). And as Stanley watches "Sylvia angling in the Brook," he beholds "the fishes strife, / Which first should sacrifice its life, / To be the Trophy of her Hook" (lines 97–100).[7] In images of the male angler, the making of the fishing tools and the careful observation of seasons, times, and places to fish are paramount to his success—in both catching fish and manifesting the appropriate spiritual qualities. In contrast, the beauty and charm of the woman are all she needs to catch fish and to fulfill her role as a successful angler. The walk of the woman amid nature and her manipulation of the angling tools become ways through which her beauty may be admired.

In other piscatory poems, the speaker posits parts of the woman's body as the angling tools themselves. In these poems, the author bypasses the pretense of the woman's actual "angling," such that she becomes explicitly the "bait." In the course of Donne's "The Baite," for example, the speaker moves away from the idea of using the "lines . . . and hookes," and asserts to the woman that: "thee, thou needst no such deceit, / For thou thy selfe art thine owne bait" (lines 4; 25–26). The fish likewise give themselves to Lovelace's "Faire Aramantha": "What needs she other bait or charm / But look? or Angle, but her arm?" (lines 141–42).[8] As these poets specifically replace the fishing implements with the woman's body, they mark a distinct move toward the allegorizing of the image. The angling woman has become, through the male gaze, herself the "angle" and the "bait"— not for the fish alone, but also for the speaker himself.

And thus the image of the fish swimming willingly to their deaths upon the woman's hook modulates allegorically to that of the man finding himself hopelessly caught by her beauty and charm. Here, the symbolic inversion of the dominant piscatory is achieved as the caught "man" becomes specifically the "male." In Waller's "Upon a Lady's Fishing," Clorinda catches "whole shoals of men"; more pointedly for the speaker, she also catches his "heart" (lines 36, 38). In another poem by Waller, "On St. James's Park," "The ladies, angling in the crystal lake" are "At once victorious with their lines, and eyes, / They make the fishes, and the men, their prize" (lines 33; 35–36). For Donne, too, "the fish that is not catched thereby [by the "bait" of her body], / Alas, is wiser farre then I" (28).[9] In a few piscatory poems of the feminine, the speaker begrudges the fish who are allowed to see the body of the bathing woman while the speaker himself must remain "uncaught"[10]; but the majority of speakers in these poems express a sense of sportive resistance to the woman's successful "fishing."

Through her successful angling of men, the woman angler completes her allegorical variation on the piscatory pastoral and finds the place within this genre that she will maintain through most of the seventeenth century. Men throughout the century enjoy full realistic representation, for proper instruction in the details of angling becomes a key element in exploring the allegory of the "spiritual angler," whose life is well-ordered and skillfully lived. The woman's angling, on the other hand, receives little detailed description, for a description of her beauty alone suffices to sustain the allegory of sexual attraction and physical love. The representation of the actual woman angler, then, is both defined and limited by the parameters of her allegorical value within the piscatory tradition.

IV

Only at the end of the seventeenth century does the woman find reinstatement in the mainstream of the British piscatory. In these later georgic piscatory poems, Waller's pattern for the representation of the woman angler is replaced by a more realistic model, in which the woman is given equal footing with the man and her beauty ceases to be her most prominent feature. In the *Innocent Epicure* (1697), for example, the anonymous male author speaks of himself and his wife who together "Haste to the neighbor Streams our luck to try," such that even if they are "baulk'd in Sport," they "return assur'd of Joy" (63). Here, the game of love is not an outgrowth of the allegorizing of the woman angler; rather, it is a mutual enterprise, independent of success or failure in fishing.

This shift in the feminine piscatory accompanied a general decline in the use of allegory and an increased attention to clarity of expression and realistic representations of labor in later seventeenth-century British literature; thus, the trappings of allegory tended to fall away as fishing treatises and georgic poems proliferated.[11]

Furthermore, the rise of the natural sciences in the later seventeenth century brought about a gradual shift from an anthropomorphic view of nature—in which nature was seen to respond sympathetically to human behaviors and qualities (such that all of nature, including the fish, submits to the woman's beauty)—to a view that saw nature as an independent entity, to be explored, studied, and understood (such that the woman angler is represented in actual contact, and contest, with her natural surroundings).[12] Although the limited condition of women did not change significantly during this period,[13] a space did exist in the 1690s for such an adventurous woman as Celia Fiennes to travel through England virtually alone, stopping to fish along the way. And the focus of poetry about the angling woman, freed from her allegorical constraints and from the assumption that her success depended on nature's responding to her beauty, could now turn to an examination of her skill in wielding the angling tools and catching actual fish.

In his long, rambling georgic, "The Genteel Recreation" (1700), John Whitney at last gives expression to the actual angling of an autonomous woman. In this poem, he depicts in verse a woman as an experienced angler, capable of making a solitary excursion and of fishing with skill: "A Reverend Matron [Mrs. Bruges, of Withyham] with a Hook and Line, / Had nick'd the most auspicious time" (17). Rather than being adored by nature, this woman observes her natural surroundings and chooses the proper time for angling. The poet continues, describing the woman's techniques:

> Silent she goes and takes a shady stand,
> Watchful her eye and steady was her hand,
> For well she knew them both for to command,
> A worm well scour'd without the help of stinking tar,
> That was her bait and that was best by far. (17–18)

In contrast to Waller or Donne, Whitney gives a detailed account of the woman's skill and knowledge of the sport. She is "silent," "watchful," "steady," and in "command." She chooses her bait wisely (indeed, "worms" are notably absent from earlier woman angler poems). As in the dominant (male) piscatory of the seventeenth century, here the woman's skill rather than her beauty is emphasized; indeed, her beauty does not even receive mention. And in the end, "My Matron at the Fishing Plot . . . packs up her Tools and homeward goes / Well Laden with a Brace or more, / The just expense of but one only hour" (18). In Whitney's poem, the woman achieves what men throughout the seventeenth century had been shown accomplishing, and what women had been denied: she returns home not with a man, but rather with a brace of fish.

But Mrs. Burges is a relative latecomer on the "woman angler" scene. The paradigm of the capable woman angler fishing knowledgeably, successfully, contemplatively, alone—the model which to this day favorably persists—was slow in replacing that of Waller's Clorinda. Through most of the seventeenth century, the "holsom walke and mery" of Juliana Berners, the delightful fishing of Margaret Hoby or Hannah Woolley or Celia Fiennes, seem lost upon male poets. Only at the end of the century, through the pens of such poets as Whitney, do the Clorindas and Sylvias and Aramanthas of the seventeenth-century piscatory begin to receive credit not only for their beauty and success in love, but also for their skill and success in fishing.

Notes

1. See, for example, Paterson; Morris, *Uncommon Waters*; Morris, *Different Angle*; and Foggia.
2. See, for example, Letcher, Eclogue IV, stanzas 5–12; 27–31 (192–98). See also Walton, as well as Bevan, chapter 3, on this aspect of Walton.
3. For discussions of Berners as author of the *Treatise*, see Duggan, McDonald, Hands, and Braekman.
4. For an example of these elements in a sixteenth-century treatise, see the anonymous *The Arte of Angling*.
5. See also Drayton, whose speaker asserts that "Oft have I seene the Sunne / To doe her [Sirena] honour / Fix[ing] himselfe at his noone, / To look upon her" (lines 216–19).
6. See also Donne's "The Baite," in which the speaker begins by asserting that he and the woman will venture out "With silken lines, and silver hookes" (line 4).
7. Similarly, in Drayton's poem, the fish "stive a good / Them to entangle" on the hook of Sirena; or even "leaping on the Land . . . Their Scales upon the sand, / Lavishly scatter" so that she may see herself as if in a mirror (lines 252–57).
8. Some poems omit any reference to the fishing tools and merely depict the trope of the fish sacrificing itself to the woman. See, for example, an anonymous parody of Marlowe's "The Passionate Shepherd to his Love," in the 1600s *England's Helicon*, a poem in

which the shepherd asserts to his lady that, as she rests in a tree above the river, she will see "the fishes gliding on the sand: / Offring their bellies to your hands" ("Another of the same . . . ," lines 15–16). Similarly, in William Browne's *Britannia's Pastorals* (1625), the water god sings of the "pure" Marine that "The best Fishes in my flood / Shall give themselves to be her food" (book 1, song 2, lines 53–4).

9. See Cunnar for an extensive discussion of this image in Donne's poem.

10. For example, in his "The Request" (1647), Cowley asks: "What service can mute fishes do to thee? / Yet against them thy dart prevails, . . . Dost thou deny only to me / The no-great privilege of captivity?" (lines 41–2; 45–6).

11. For a discussion of this decline in the use of allegory in the course of the seventeenth century, see Murrin.

12. Thomas gives an insightful and well-documented account of changes in the perception of nature in early modern England; see esp. "Changing Perspectives," 87–91.

13. See Fraser for an extensive discussion of the condition of women in seventeenth-century England; for a brief account of the lack of progress in this area in the course of the century, see esp. 464–70.

Works Cited

"Another of the same nature [as 'The passionate Sheepheard to his love'], made since." *England's Helicon 1609, 1614.* Ed. Hyder Edward Rollins. Cambridge: Harvard UP, 1935. I: 187–88.

The Arte of Angling (1577). Ed. Gerald Eades Bentley. Princeton: Princeton UP, 1958.

BERNERS, DAME JULIANA. *The Treatise on Angling in "The Boke of St. Albans"* (1496). Ed. W. L. Braekman. Brussels: Scripta, 1980. 56–83.

BEVAN, JONQUIL. *Izaak Walton's* The Compleat Angler: *The Art of Recreation.* Sussex: Harvester, 1988. 67–99.

BRAEKMAN, W. L. *The Treatise on Angling in "The Boke of St. Albans (1496)": Background, Context and Text of* The Treatyse of Fysshynge wyth an Angle. By Dame Juliana Berners. Brussels: Scripta, 1980.

BROWNE, WILLIAM. *Britannia's Pastorals. Poems of William Browne of Tavistock.* 2 vols. Ed. Gordon Goodwin. London: Routledge, 1971.

COWLEY, CABRAHAM. "The Request." *The Poems of Abraham Cowley.* 3 vols. Chiswick, 1822. 2: 5–7.

CUNNAR, EUGENE R. "Donne's Witty Theory of Atonement in 'The Baite.'" *Studies in English Literature* 29 (1989): 77–98.

DENNYS, JOHN. *The Secrets of Angling: Teaching, The choisest Tooles Baytes and seasons, for the taking of any Fish, in Pond or River: practised and familiarly opened in three Bookes.* London, 1613.

DONNE, JOHN. "The Baite." *John Donne: The Complete English Poems.* Ed. A. J. Smith. London: Penguin, 1971. 43–44.

DRAYTON, MICHAEL. "The Sheperds Sirena." *Minor Poems of Michael Drayton.* Ed. Cyril Brett. New York: Books for Libraries, 1972. 151–60.

DUGGAN, ALFRED. "The Lady and the Trout, Part I: The Writing of the Treatise." *Sports Illustrated,* 13 May 1957: 75–87.

FIENNES, CELIA. *The Illustrated Journeys of Celia Fiennes.* London: Macdonald, 1982.

FLETCHER, PHINEAS. *Piscatorie Eclogues* (1633). *Giles and Phineas Fletcher: Poetical Works.* 2 vols. Ed. Frederick S. Boas. Cambridge: Cambridge UP, 1909. 2: 175–222.

FOGGIA, LYLIA. *Reel Women: The World of Women Who Fish.* Hillsboro: Beyond Words, 1995.

FRASER, ANTONIA. *The Weaker Vessel: Woman's Lot in Seventeenth-Century England.* London: Weidenfeld and Nicolson, 1984.

HANDS, R. "Juliana Berners and the Boke of St. Albans." *RES* 18 (1967): 373–86.

HOBY, LADY MARGARET. *Diary of Lady Margaret Hoby.* Ed. Dorothy M. Meads. Boston: Houghton Mifflin, 1930.

Innocent Epicure. London, 1697.

LOVELACE, RICHARD. "Aramantha: A Pastoral" (1649). *The Poems of Richard Lovelace.* Ed. C. H. Wilkinson. Oxford: Clarendon, 1930. 107–18.

MARKHAM, GERVASE. *The Pleasures of Princes, or Good mens Recreations: Containing a Discourse of the generall Art of Fishing with the Angle, or otherwise. and of all the hidden secrets belonging thereunto. (London, 1614). Three Books on Fishing.* Ed. J. Milton French. Gainesville: Scholar's Facsimiles and Reprints, 1962.

MCDONALD, JOHN. *The Origins of Angling.* Garden City: Doubleday, 1963.

MORRIS, HOLLY, ed. *A Different Angle: Fly Fishing Stories by Women.* Seattle: Seal, 1991.

———. *Uncommon Waters: Women Write about Fishing*. Seattle: Seal, 1991.

MURRIN, MICHAEL. "The End of Allegory." *The Veil of Allegory: Some Notes Toward a Theory of Allegorical Rhetoric in the English Renaissance*. Chicago: U of Chicago P, 1969. 167–98.

PATERSON, WILMA, AND PETER BEHAN. *Salmon and Women: The Feminine Angle*. London: H. F. & G. Witherby, 1990.

STANLEY, THOMAS. "Sylvia's Park" (1651). *The Poems and Translations of Thomas Stanley*. Ed. Galbraith Miller Crump. Oxford: Clarendon, 1962. 156–63.

THOMAS, KEITH. *Man and the Natural World: A History of the Modern Sensibility*. New York: Pantheon, 1983.

WALLER, EDMUND. "On St. James's Park." *The Poems of Edmund Waller*. Ed. G. Thorn Drury. New York: Greenwood, 1968. 168–73.

———. "Upon a Lady's Fishing with an Angle." *The Poems of Edmund Waller*. Ed. G. Thorn Drury. New York: Greenwood, 1968. 244–45.

WALTON, IZAAK. *The Compleat Angler: 1653–1676*. Ed. Jonquil Bevan. Oxford: Clarendon, 1983.

WHITNEY, JOHN. *The Genteel Recreation: or, the Pleasure of Angling, a Poem. With a Dialogue between Piscator and Corydon*. London, 1700.

WOOLLEY, HANNAH. *The Accomplish'd Lady's Delight. In Preserving, Physick, Beautifying and Cookery*. London, 1675.

Dorothy Wordsworth, Ecology, and the Picturesque

ROBERT MELLIN

William Wordsworth presented "a harmonious relationship with nature," claims Jonathan Bate in *Romantic Ecology*, that "goes beyond, in many ways goes deeper than, the . . . model we have become used to thinking with. By recapturing the Wordsworthian pastoral, we may begin to reconfigure the model" (19–20). Although Bate's premise should intrigue anyone working toward an environmentally sound transformation of culture, his inattentiveness to gender and the literary marketplace of late-eighteenth century England problematizes his concept of romantic ecology. Recent scholarship has shown that women writers and women readers dominated the literary marketplace when *Lyrical Ballads* was first published, and suggests the incompleteness of a study that is not informed by some consideration of the extent to which the presences of women readers and writers circumscribed the construction of "Wordsworthian" principles. One woman in particular, Dorothy Wordsworth, certainly influenced this "romantic ecology": her picturesque written descriptions were often appropriated by William without attribution. In his study, Bate does pause to note how Wordsworth's "eye for the detailed observation of nature was opened by his sister" (39), but this is as far as Bate goes with his consideration of female influence. Oddly enough, then, Bate perpetuates a tradition begun by William Wordsworth himself in *Lyrical Ballads* by erasing Dorothy's influence from the text.¹ In fact, Dorothy Wordsworth's picturesque writings anticipate the nonanthropocentric, bioregionalist efforts of many environmental writers today. Thus, a truly radical attempt "to reconfigure" our "model" of a harmonious relationship with nature relies on both contextualizing the origins of William's work within the markeplace of the 1790s and reconsidering Dorothy Wordsworth's thinking and writings. Such a reconsideration suggests that Dorothy Wordsworth's use of the picturesque brings to romantic ecology a material specificity often lacking in William's more abstract representations.

The preeminence of women in the literary world of late eighteenth-century England, a fact until recently suppressed in most literary histories, was "universally known among the literate of the 1790s," writes Stuart Curran (186). There were "many more women than men novelists," and "the theater was actually dominated by women. . . . In the arena of poetry . . . the place of women was likewise, at least for a time, predominant" (Curran 186–87). A sampling of that era's texts, which are now being made more readily available by the publication of anthologies such as Jennifer Breen's *Women Romantic Poets: 1785–1832*, indicates that William Wordsworth's claims to originality in the "Preface" to *Lyrical Ballads* may actually have been an attempt to deny the existence of his predecessors.² As Breen points out, the "precise descriptions of the natural phenomena and yeomen people" found in Joanna Baillie's "A Summer's Day" and "A Winter's Day" preceded Wordsworth's similar work by nearly a decade (xxii). Charlotte Smith's poetry resonates even more obviously with the style and subject matter later associated with Wordsworth. Curran, in noting the "long, sinuous verse paragraphs, the weighted monosyllables, [and] the quick evocation of natural detail" in Smith's poetry, concludes that Smith's 1793 *The Emigrants* contributed significantly to Wordsworth's development (202). Yet Wordsworth chose to suppress his indebtedness, a decision apparently motivated by economic and by personal reasons.

An appeal to the literary marketplace of the 1790s, which underwrote the production of *Lyrical Ballads*, was prompted by the need of the Wordsworths and Coleridges to overcome financial difficulties. Money from an annuity and "from the sale of a volume to be called *Lyrical Ballads*," Amanda Ellis points out, "were to finance a trip for [them] to Germany" (132–33). A letter from Dorothy Wordsworth to Mrs. William Rawson corroborates Ellis's comment. "[O]ur regular income," she writes, "will be sufficient to support us when we are there, and we shall receive, before our departure much more than sufficient to defray the expenses of our journey, from a bookseller to whom William has sold some poems" (*The Letters of William and Dorothy Wordsworth: The Early Years* 224). Yet, as Breen observes, "Joanna Baillie found that her anonymously written first volume of poetry in 1790 attracted no attention" (xxii). In order to spark wider commercial interest in *Lyrical Ballads* than was generated by Baillie's volume, then, Wordsworth appears to have found it useful to inscribe the aesthetic space established by women writers as masculine.

Although women dominated the literary scene, the larger dominant culture of 1790s England was more likely to respond to a male writer. This intersection of gender, power, and marketplace is particularly well-illustrated in a dispute between Wordsworth and Mary Robinson. Robinson, whom Curran calls "a major literary voice of the 1790s" (186), published a volume of poetry entitled *Lyrical Tales* in 1800, the same year the second edition of *Lyrical Ballads* was published. Dorothy Wordsworth writes that she and William found Robinson's title "a great objection" because they already had claimed the title with the publication of *Lyrical Ballads* two years earlier (*Letters: The Early Years* 297). That Wordsworth could object to the use of "his title" when he was, in fact, coopting the aesthetic space Robinson helped create, reveals the speed with which he moved from exploiting the market created by women to claiming that market as his own.

Personal interests also motivated William Wordsworth to hide his indebtedness to women writers. As Breen notes, Anne Grant's "celebration of the domestic foreshadow[ed] Wordsworth's recognition that the experience of ordinary people can be an eminently suitable subject for poetry" (xvii); Wordsworth, however, writes in the 1802 "Preface" to *Lyrical Ballads*, "Poetry is the most philosophic of all writing: it is so: its object is truth, not individual and local, but general, and operative" (605). Wordsworth's poetics dictated that he work the "local" truths of the domestic sphere into "philosophic" and "general" abstractions about ordinary people. The 1802 "Preface" exemplifies how Wordsworth, in the four years following the first publication of *Lyrical Ballads*, had become uncomfortable with his bifurcated poetics. In the "What is a Poet?" passage, Wordsworth repeats ad nauseam that the Poet is a "man." This "man" is gender-specific and not the more general reference to mankind or humanity, for as Wordsworth begins "What is a Poet?" he argues that it "might appear to some" that "I am like a man fighting. . . " (603). The Poet, moreover, "is a man speaking to men: a man, it is true. . . ." (603). In fact, Wordsworth points out that any slip in the manly battle of poesy will lead to disgraceful "unmanly despair" (604). Karl Kroeber's suggestion that Wordsworth was "profoundly secure in his masculinity" (136) is therefore a bit misleading: Wordsworth's need to be masculine was itself generated from his uneasiness with the domestic sphere.

William's need to appropriate and efface the writing done by women during the late eighteenth century complicated his relationship with Dorothy. Dorothy's journals were of great interest to William, but since journal writing occupied the feminized domestic sphere in William's sexual division of poetics, he was forced to suppress Dorothy's voice in his texts. Dorothy was aware of William's position: she wrote, "Journals we shall have in sufficient number to fill a Lady's bookshelf,—for all, except my Brother, write a Journal" (*Letters: The Middle Years* 625). The Poet, of course, should not write journals. Some of Dorothy's journal writing nonetheless did find its way into William's poems. William uproots the "daffodils" entry from Dorothy's *The Grasmere Journals* (*Grasmere* 109) and transplants it into the second stanza of "I wandered lonely as a cloud." Similarly, parts of Dorothy's 4 February 1798 entry in *The Alfoxden Journal* appear in lines 742–45 of William's "The Ruined Cottage" (*Alfoxden* 5).[3] Yet, as Susan Wolfson notes, this arrangement caused a "quiet agitation" in William (147). For example, William becomes upset with Dorothy when she reads him an "account of the little Boys belonging to the tall woman"; the reading was "an unlucky thing," Dorothy writes, "for he could not escape from those very words, and so he could not write the poem" (*Grasmere* 101). He does overcome his distress, however, and places Dorothy's original account of "a very tall woman" (*Journals of Dorothy Wordsworth* 26) into the opening lines of "Beggars." William's "quiet agitation" also caused him to represent "experiences [William and Dorothy] shared as solitary ones" (Wolfson 147), as may be seen in William's deletion of a lengthy passage concerning Dorothy from his 1798 "Nutting." The "Nutting" erasure suggests that William contemplated, but rejected, a more inclusive poetics.

The ecological usefulness of William's nature writing is therefore limited, because as he distanced himself from Dorothy's writing he also distanced himself from picturesque writing. The picturesque, which Nicholas Roe calls "a way of seeing that reconciles humankind with nature" (120), was Dorothy's preferred style of writing in her journals, and it appealed to the increasingly large middle-class female readership. Although William Snyder claims that William left behind the picturesque "to seek the divine in Nature" (148), that abandonment has a more material basis than Snyder suggests: William abandoned the picturesque in order to distance himself from "women's writing." William's gender politics, if Roe's observation is correct, call into question just how harmonious his relationship with nature can be. What is needed in order to construct a more productive Wordsworthian ecology, therefore, is a reconsideration of Dorothy Wordsworth's picturesque representations of nature.

The Alfoxden Journal provides the most intriguing examples of Dorothy's use of the picturesque as ecologically informed writing. These 1798 entries were written at a time when she saw herself as a writer sharing in the observations of nature with Coleridge and William (and just as William was preparing to distance himself from her). In the more frequently cited *The Grasmere Journals*, Dorothy acted more as a compiler of "data . . . for William" (Fay 126) than as a writer. Her position in *The Grasmere Journals* thus leads John Nabholtz to claim that since the landscape at Grasmere provided better "pictorial effects" (121) than did the landscape at Alfoxden, Dorothy's picturesque writing is more vivid in *The Grasmere Journals*, and it is therefore her strongest work. *The Alfoxden Journal*, however, does not simply provide picturesque fodder for William's use; in it, she uses the picturesque in order to gain a sense of place, which in turn allows her to represent the interconnectedness of humans and the environment more clearly than did William. Nonetheless, the temptation to read *The Alfoxden Journal* as simply an exercise in the picturesque is understandable. For example, in her first entry, dated 20 January 1798, Dorothy writes:

> The green paths down the hillsides are channels for streams. The young wheat is streaked by silver lines of water running between the ridges, the sheep are gathered together on the slopes. After the wet dark days, the country seems more populous. It peoples itself in the sunbeams. The garden, mimic of spring, is gay with flowers. The purple-starred hepatica spreads itself in the sun, and the clustering snow-drops put forth their white heads, at first upright, ribbed with green, and like a rosebud; when completely opened, hanging their heads downwards, but slowly lengthening their slender stems. The slanting woods

of an unvarying brown, showing the light through the thin network of their upper boughs. Upon the highest ridge of the round hill covered with planted oaks, the shafts of the trees show in the light like the columns of a ruin. (1)

The passage strictly adheres to the principles of picturesque representation, which is not to say, however, that *The Alfoxden Journal* itself is thus an exercise in the picturesque. In "The Structure of the Picturesque: Dorothy Wordsworth's Journals," Robert Con Davis painstakingly lists each of numerous picturesque aspects of this journal entry.[4] It is, as he notes, a "full visual composition" in which the "foreground-and-background organization establishes a fixed perspective and position" (45). But by focusing on only one example from *The Alfoxden Journal*—a dismissive gesture generated from the notion that Dorothy Wordsworth's importance is based on her ancillary descriptive contributions to William's works—Davis draws the poorly supported conclusion that "Dorothy uses landscape imagery to control nature through a highly rational method of comprehending the world" (46). Leaving aside the questionable underlying assumption that there is indeed an unmediated way of "comprehending the world," it is uncertain whether Dorothy "used" the picturesque to control nature. If anything, the picturesque is a "highly rational method" of representation that allows Dorothy to position herself nonanthropocentrically.

Dorothy's ecologically sound critical position becomes more obvious as one reads past the initial entry. After the 20 January entry, Dorothy spends less time honoring the formal structure of the picturesque and more time representing the physical structure of the Alfoxden region. Her representation of place is bioregionalist. A bioregion "can be identified by its mountain ranges and rivers, its vegetation, weather patterns or soil types, or its patterns of animal habitats, whether birds, ground mammals or humans" (Tokar 27)—precisely the features Dorothy identifies in her journals. On 22 January, for example, she describes "hollies, capriciously bearing berries" and then writes: "Query: Are the male and female flowers on separate trees?" (1). Similarly, on 5 February she writes: "Observed some trees putting out red shoots. Query: What trees are they?" (5). Her writing demonstrates that Dorothy is engaged in learning about, rather than controlling, the Alfoxden bioregion. In order to gain a sense of place, she observes and describes phases of the moon, positions of Jupiter and Venus, weather patterns, birds, vegetation, rivers, and patterns of animal and human behavior. Dorothy, along with Coleridge, also searches out the region's watershed. On 6 April she writes that they walked with "an intention to find the source of the brook," and on 7 April she notes that she "Walked . . . to the source of the brook" (12). She is so acutely aware of her surroundings that she even notes subtle absences: "The sound of the sea distinctly heard on the tops of the hills, which we could never hear in summer. We attribute this partly to the bareness of the trees, but chiefly to the absence of the singing of birds, the hum of insects, that noiseless noise which lives in the summer" (2). Although John Nabholtz argues that Dorothy's writing provides a "means of discovering and describing the *inherent* beauty of nature" (128), her journal's entries actually focus on the inherent *structure* of the Alfoxden ecosystem.

Dorothy's de-idealized reading of the Alfoxden region also heightens her awareness of human/nonhuman relations. Davis finds her representations of humans in the landscape as a shortcoming, though. He writes, "Experience can be subordinated so fully to picturesque description that landscape figures may appear and fade as they are needed for the composition," and "people easily get lost in the foliage where human relationships are absent," becoming "discrete shadows" (47). Dorothy's supposed eschewal of "the bothersome complexities of personal relationships" (Davis 47), however, is actually a protoecological decentering of human representation in the overall physical landscape. Humans aren't "absent" in the foliage, but are represented as integrated into their surroundings, a nonanthropocentric representation of humans that seems increasingly useful to twentieth-century environmental writers. The first human does not appear in *The Alfoxden Journal* until her 26 January entry, and that mention is a quick reference to a "woodman" (2). Moreover, on 10 March she describes "groups of human creatures, the young frisking and dancing in the sun," and on 11 March she comments that it was "Pleasant to see the labourer on Sunday jump with the friskiness of a cow upon a sunny day" (10). The distinctions between human and animal representations blur. Her careful bioregionalist reading of the Alfoxden region contributes to her careful reading of the place humans have within the structure of the region, which is in marked contrast to the anthropocentric representation of nature in traditional environmental writing. Dorothy's "local" truths aren't used to make "general" truths, but instead they have their own intrinsic value.

Dorothy's materialist reading of the region and of the relationship of humans and the environment lead her toward a disconcerting recognition of gender relations at the end of eighteenth-century England. In her reading of the landscape, she notes that men have a privileged place in the social structure. She describes one of the men in the landscape, for example, as "a razor-grinder with a soldier's jacket on, a knapsack on his back, and a boy to drag his wheel" (7). Women, however, fit into the landscape differently. On 4 February, for example, she describes "The young lasses . . . in their summer holiday clothes—pink petticoats and blue. Mothers with their children in arms. . . " (5). The contrast is striking. The razor-grinder has a place in public life, both as a worker and as a soldier. The boy, as his apprentice, is being prepared to take a place in public life. The "young lasses," however, are in their "summer holiday clothes," a description that suggests they are on holiday from work and from other modes of social production. The position of women is within the domestic sphere: they are simply "Mothers."

As Dorothy's awareness of the ecological and social processes of the region sharpens, she comes to realize what her own position is. On 22 March, Dorothy writes that she "spent the morning in starching and hanging out linen" (11); on 28 March, she "Hung out the linen" (11); and on 10 April, she notes that "I was hanging out the linen in the evening" (13). Dorothy represents herself in the landscape and in her journals as part of the domestic scene. Her careful reading of the landscape causes her to recognize the inescapably unproductive position she occupies as a writer, a position that William is on the verge of exploiting. Beginning with the 22 March entry, and becoming more pronounced after 10 April, Dorothy's journal entries lose much of the rich detail of her earlier picturesque descriptions. She seems to resign the position she had developed earlier in the journal.

In fact, after 14 April she appears to abandon her public life as a writer. On April 14, she writes: "Walked in the wood in the morning. The evening very stormy, so we staid within doors. Mary Wollstonecraft's life, etc., came" (13). Godwin's biography, which brought unexpected disrepute to Wollstonecraft, may have ruined Dorothy's already faltering efforts. Having heard this morality play on Wollstonecraft's life, after having seen the social space women must occupy in the cultural landscape, Dorothy seems hesitant to occupy the public position that Wollstonecraft did, and for which she was scandalized—the position of a writer. The next day, 15 April, Dorothy writes that there are "Quaint falls about, about which Nature was very successfully striving to make beautiful what art had deformed" (13). In that same entry, Dorothy also writes about the "unnaturalized trees" she had noticed that day. Her earlier refusal to editorialize on the "proper" order of the bioregion has been abandoned, perhaps because of her heightened awareness of what was being constructed as the "natural" order of gender relations. Before 15 April, Dorothy used the picturesque simply to represent the ecological and social landscape of the Alfoxden region. Her didactic commentary is based on an idealized notion of nature and the natural, which is much more reminiscent of William's "general" truths than her earlier "local" ones. She evaluates the landscape, with some parts of it properly belonging and some parts not. She also seems to conclude that she does not properly belong in the public sphere. Although her earlier use of the picturesque opened her eyes to the material relations of the ecosystem and culture of the Alfoxden region, she now chooses not to see that same region, and she chooses not to be seen. Entries in *The Alfoxden Journal* lack descriptive passages after 15 April, and it quickly ends. Ultimately, as her later work in *The Grasmere Journals* illustrates, Dorothy's role as a writer becomes nearly invisible. She facilitates William's use of her descriptive writing, suggesting that she wished to avoid scandalization similar to Wollstonecraft's by adopting a nearly transparent presence during the ascendance of the "masculine" early Romantic period.

Dorothy's decision coincides with the release of *Lyrical Ballads* and with the onset of William's repudiation of women's writing. As William eclipses Dorothy's literary production, the Wordsworthian pastoral also eclipses an environmentally sensitive representation of the landscape, the picturesque. In "Floating Island," one of the few extant poems of Dorothy's, there is a description of "a slip of earth" (line 5) floating on the water. This "peopled *world*" (line 16) survives "many seasons' space" (line 16). The floating island, though, is:

> Buried beneath the glittering Lake!
> Its place no longer to be found,
> Yet the lost fragments shall remain,
> To fertilize some other ground. (lines 25–28)

Thought to be written in the 1820s, "Floating Island" illustrates that Dorothy's de-idealized perception of the environment does survive her abandonment of public writing. As Meena Alexander points out, Dorothy's "disclaimer of any right to authorship must be understood in the light of the fact that it was William who read her poems, in his voice. He publicized them. They were mediated by him. The writer had no place in the public sphere where she might have claimed her own words" (208). Dorothy's voice, "Buried beneath" William's texts and decades of male-privileged criticism, can be found in *The Alfoxden Journal*—the "lost fragments shall remain." The greening of Wordsworthian scholarship must not evade issues of gender; in fact, a careful reading of nature and gender may point to the work done by other early ecological writers and help ecocritics negotiate an ecological transformation of culture.

Notes

I wish to thank Anca Vlasopolos and members of the "Unmentionable Origins" seminar at Wayne State University for their contributions to this project.

1. Bate's "romantic ecology" is, in part, a response to Jerome McGann's *Romantic Ideology*. Bate claims that McGann's "case . . . serve[s] no constructive purpose—its only function was a destructive and local one, the breaking down of the Hartman hegemony" (10). McGann is oddly instructive, though. When he notes that "From Wordsworth's vantage, an ideology is born out of things which (literally) *cannot* be spoken of" (*Romantic Ideology* 91), he locates a key problem in Wordsworth scholarship. The ideology of Bate, Geoffrey Hartman, M. H. Abrams, and, ironically, McGann seemingly cannot speak of gender—theirs is a patriarchal ideology.

2. See Marie-Helene Huet's *Monstrous Imagination* (Cambridge: Harvard UP, 1993) for an insightful examination of masculinist claims of originality.

3. In fact, most traceable appropriations are well-documented, although I'm not aware of any single text that documents all of the known appropriations. Refer to Susan M. Levin's *Dorothy Wordsworth and Romanticism* (esp. 14), the footnotes of Mary Moorman's *Journals of Dorothy Wordsworth*, and the footnotes of nearly any scholarly edition of William's works (for example, Stephen Gill's *William Wordsworth*) for a fairly thorough list. The numerous erasures in

Dorothy's journals and the "disappearance" of *The Alfoxden Journal* manuscript limit the odds of compiling a definitive list.

4. See page 45 of his article for a convincing formal analysis of Dorothy's use of the picturesque. For a cultural analysis of this same passage, refer to Meena Alexander's "Dorothy Wordsworth: Grounds of Writing," which demonstrates that a consideration of gender can produce a radically different reading (see esp. 202).

Works Cited

ABRAMS, M. H. *The Mirror and the Lamp: Romantic Theory and Critical Tradition*. New York: Oxford UP, 1953.

ALEXANDER, MEENA. "Dorothy Wordsworth: The Grounds of Writing." *Women's Studies* 14 (1988): 195–210.

BATE, JONATHAN. *Romantic Ecology: Wordsworth and the Environmental Tradition*. New York: Routledge, 1991.

BREEN, JENNIFER, ED. Introduction. *Women Romantic Poets 1785–1832: An Anthology*. London: Dent, 1992. xi–xxviii.

CURRAN, STUART. "The I Altered." *Romanticism and Feminism* Ed. Anne K. Mellor. Bloomington: Indiana UP, 1988. 185–207.

DAVIS, ROBERT CON. "The Structure of the Picturesque: Dorothy Wordsworth's Journals." *Wordsworth Circle* 9 (1978): 45–49.

ELLIS, AMANDA M. *Rebels and Conservatives: Dorothy and William Wordsworth and Their Circle*. Bloomington: Indiana UP, 1967.

FAY, ELIZABETH. *Becoming Wordsworthian*. Amherst: U of Massachusetts P, 1995.

HARTMAN, GEOFFREY. *Beyond Formalism: Literary Essays 1958–1976*. New Haven: Yale UP, 1970.

KROEBER, KARL. "'Home at Grasmere': Ecological Holiness." *PMLA* 89 (1974): 132–41.

LEVIN, SUSAN M. *Dorothy Wordsworth and Romanticism*. New Brunswick: Rutgers, 1987.

MCGANN, JEROME J. *The Romantic Ideology: A Critical Investigation*. Chicago: U of Chicago P, 1983.

NABHOLTZ, JOHN R. "Dorothy Wordsworth and the Picturesque." *Studies in Romanticism* 3 (1963): 118–28.

ROE, NICHOLAS. *The Politics of Nature: Wordsworth and Some Contemporaries*. London: Macmillan, 1992.

SNYDER, WILLIAM C. "Mother Nature's Other Natures: Landscape in Women's Writing, 1770–1830." *Women's Studies* 21 (1992): 143–62.

TOKAR, BRIAN. *The Green Alternative: Creating an Ecological Future*. San Pedro: Miles, 1987.

WOLFSON, SUSAN J. "Individual in Community: Dorothy Wordsworth in Conversation with William." *Romanticism and Feminism*. Ed. Anne K. Mellor. Bloomington: Indiana UP, 1988. 139–66.

WORDSWORTH, DOROTHY. *The Alfoxden Journals. Journals of Dorothy Wordsworth* Ed. Mary Moorman. New York: Oxford UP, 1971. 1–14.

———. "Floating Island." *Women Romantic Poets 1785–1832: An Anthology*. Ed. Jennifer Breen. London: Dent, 1992. 131–32.

———. *The Grasmere Journals. Journals of Dorothy Wordsworth*. Ed. Mary Moorman. New York: Oxford UP, 1971. 15–166.

———. *Journals of Dorothy Wordsworth*. Ed. Mary Moorman. New York: Oxford UP, 1971.

WORDSWORTH, DOROTHY, AND WILLIAM WORDSWORTH. *The Letters of William and Dorothy Wordsworth: The Early Years, 1787–1805*. 2nd ed. Ed. Ernest De Selincourt. Rev. Chester L. Shaver. Oxford: Oxford UP, 1967.

———. *The Letters of William and Dorothy Wordsworth: The Middle Years, Part II, 1812–1820*. 2d. ed. Ed. Ernest Deselincourt. Rev. Mary Moorman and Alan G. Hill. Oxford: Clarendon, 1970.

WORDSWORTH, WILLIAM. "Beggars." *William Wordsworth*. Ed. Stephen Gill. New York: Oxford UP, 1984. 243–44.

———. "I Wandered Lonely as a Cloud." *The Columbia Anthology of British Poetry*. Ed. Carl Woodring and James Shapiro. New York: Columbia UP, 1995. 441.

———. "Preface to *Lyrical Ballads, with Pastoral and Other Poems* (1802)." *William Wordsworth*. Ed. Stephen Gill. New York: Oxford UP, 1984. 31–44.

Mary Austin's Nature
Refiguring Tradition through the Voices of Identity
ANNA CAREW-MILLER

Mary Austin's notoriety died with her in 1934, and her writings, well known during her lifetime, remained relatively obscure until recent years. Much of Austin scholarship has focused on her significant contribution to a women's tradition of nature writing and, in particular, on her proto-feminist characterizations of the natural world. Austin's life has received as much attention as her work; many critics read her work as a revelation of her struggles as a woman writer at the turn of the century.[1] Biographical criticism of this sort purports to look for the *real* Mary Austin, the truth about her life. However, I believe richer readings of Austin's work can be attained through persona criticism, which is concerned not so much with biography as with reading the literary self-constructions of the persona, or voice, a writer uses to narrate his or her work.[2] The persona or narrative voice is created by a writer's psychology and history but doesn't add up to a historical figure. Rather than read Austin's work for information about her struggles as a woman or for insight into her contributions to the field of nature writing, I propose to read Austin's narrative voice in order to arrive at a clearer understanding of how her representations of nature reveal her sense of place within the natural world. I want to explore how her narrative voice reflects her experience of nonhuman nature.

Austin scholars have frequently made the connection between Austin's writing and her relationship with her mother, finding it to be a damper on her creativity.³ I will argue instead that Austin's struggle to define herself against her mother and the standards of femininity her mother represented gave Austin's writing its power and purpose, giving her narrative voice the idiosyncratic subjectivity that readers have found so compelling and unusual in the genre of nature writing. Rather than attempt to psychoanalyze Austin with details from her biography, we can use her biographical information to examine the inflections in Austin's narrative voice—to see, on the one hand, how the pain in Austin's life wrote through her and, on the other hand, how she used her life experiences to textually represent nature in a new way. Only by understanding how Austin's life shaped her literary methodology will we begin to understand her innovations in the genre of nature writing.

What most interests me in Austin's nature writing is the tension that is produced by the dichotomy in Austin's narrative voice. Austin's voice can be read as *two* voices, one that confidently overturns tradition and believes in her intuitive connection to nature, and one—which Austin works at rejecting—that feels trapped by traditions of Victorian womanhood and fears the natural world. What gives her writing its power is the pull between these two narrative voices. In her autobiography, *Earth Horizon* (1932), Austin labeled these voices "I-Mary" and "Mary-by-herself," respectively. I-Mary is Austin's voice of public performance, confidently narrating most of her published work. Mary-by-herself is the private voice of Austin's journals, poetry, and letters, which also appears as a shadow in her published works. Mary-by-herself speaks for Austin's fearfulness, and I-Mary speaks for her complicated desires. With I-Mary's voice, Austin successfully feminizes the masculine genre of nature writing by claiming that intuition is as valid a means of understanding the natural world as observation, by turning the tropes of "Mother Nature" into powerful goddesses, and by inventing a female character who embodies an ideal relationship to the natural world. However, I-Mary's voice is shadowed by that of Mary-by-herself, and through reading this voice, we are led to the source of her creative energy: her painful relationship with her mother.

Austin's mother, Susannah Hunter, kept her emotional distance; she was critical of her daughter's attempts to be a wife and mother and made little effort to support Austin's career as a writer.⁴ During her mother's lifetime, Austin's fear of rejection by her mother appears to have had an enervating effect, limiting her creativity and ambition. This fear is transcribed by the voice of Mary-by-herself into ominous representations of a female nature, which are found most often in her private papers. The fear of some nameless threat, a brooding presence that Mary-by-herself experiences over and over again, mimics Austin's own fear of repeated rejection by her mother. Moreover, just as Austin returned again and again to her mother, hoping her mother's response to her would change, Mary-by-herself repeatedly returns to nature, in spite of her fear. However, throughout Austin's writing career, her longing for connection with her mother, so tied to her fear of rejection, is slowly transformed into I-Mary's desire for communion with nature.

Austin pushed the voice of Mary-by-herself to the periphery of her writing as she put more distance between her own life and the world of her mother. With her marriage in 1891 and birth of her daughter in 1892, Austin seems to have separated herself from her mother enough to begin writing. Her first published story, "The Mother of Felipe," appeared in *The Overland Monthly* in 1892. Austin recalled her early work in the later *Earth Horizon*: "It was as I-Mary walking a log over the creek, that Mary-by-herself could not have managed, that I wrote two slender little sketches" (231). Simultaneously, she grew in confidence as a writer as she became more successful, and the voice of I-Mary pulled more strongly against the fears of Mary-by-herself. Austin produced her most powerful nature writing during the period between Susannah Hunter's death and Austin's own relocation from the foothills of the Sierra Nevada to Carmel, California—roughly 1896 to 1906. This writing is marked by the tension between the energies of the newly liberated voice of I-Mary and the nearly vanquished doubts of Mary-by-herself.

During this period, Austin transformed her fear into desire and used it as a creative force; seven years after her mother's death, her first book, *The Land of Little Rain*, was published. We can hear I-Mary as the knowledgeable voice in *The Land of Little Rain*, the voice of union with the landscape, of collectivity and creativity, the voice of desire for connection. Desire, for I-Mary, is not predicated on emptiness and hunger, but on abundant creative energy and appetite for connection. "Far from being detached," she writes in *Everyman's Genius*, "I have so excellent an appetite for life that if there are any experiences that a woman, remaining within the law and a reasonable margin of respectability, might have, I am still hopeful of being able to compass them" (184). This is the authentic voice of the woman unbound by social restrictions, the voice of subjective authority, hungry for experience and privileged knowledge.

In this first book, *The Land of Little Rain* (1903), Austin crafted a textual persona much different from those typically found in nature writing texts; I-Mary speaks in a highly personal voice and shares an extremely subjective vision of the desert landscape. This voice also narrates several of Austin's later works, including *Lost Borders* (1909) and *The Flock* (1906). In these later works, Austin continued to develop an alternative woman's tradition of nature writing by transforming the conventional metaphors for a female nature.[5] She replaces the conventional tropes—the bounteous mother earth and seductive nymph—with a wilder goddess, frequently representing the landscape of the desert as a lioness. In Austin's writing, this version of a female nature is powerful and dangerous. Only women who have rejected Victorian codes of conduct for ladies can live unmolested by this lioness because their relationship to nature is one of connection and communion. Austin created a character that embodied these ideals, the chisera, who provides Austin with a model for herself as a woman writer. The chisera has an intuitive knowledge of nature and a constant desire for fulfillment through close interaction with the natural world.

Austin could not always successfully shed her identity as a Victorian lady in order to become her more idealized self in the desert, particularly as a young woman who hadn't yet become a successful writer. But before examining Austin's transformative refigurations of the wild, we should listen to the voice of Mary-by-herself, through which we can hear fear in Austin's response to the wild. Mary-by-herself is diminished by the land, rendered helpless by her terrors. Melody Graulich observes: "Mary-by-herself is partially the poignant offspring of Austin's conflicted relationship with her mother, partially the outcast from social constructions of femininity that left her feeling as if she failed as a woman" (379). Perhaps this is the voice that feels the restrictions of Victorian ladyhood most severely. Austin recalls in her unpublished essay, "The Friend in the Wood": "Looking back, I realize that those years in which fear hung on the fringes of my consciousness and occasionally became a factor in all outdoor experience, coincided with the years in which I had lived most actively in my life as a woman" (194). These were her years in the land of little rain, in which she negotiated the end of her painful relationship with her mother, as well as a troubled marriage and her own trying experiences as the mother of a mentally disabled child.[6]

We first hear the fear this version of a fierce nature evoked in the journals—now called the *Tejon Notebook*—that she kept during the first few years when she lived with her mother and brother in Kern County. There she wrote of the land as sometimes having a Gothic severity and grim quality: "There are times when everything seems to have a sinister kind of life. It shows its teeth to me. At other times, it is merely beautiful and gentle." This image begins to be more clearly fleshed out in later writing. Austin's fear of the wild emerges when we hear Mary-by-herself revealing that her longing for connection and communion to the land is met by a lioness whose indifference and ferocity frighten her. Her unpublished poem "Inyo" includes a fierce rendition of the lioness, in which the "untamed and barren" valley is "Like some great lioness beside the river / With furies flaming in her half-shut eyes." In Austin's work, the lioness appears only in connection to this particular landscape—the high desert and mountains, where Austin lived during the era that the voice of Mary-by-herself was strongest.

Because Austin worked to control her fear while living in the mountains and desert of California, finding evidence of Mary-by-herself is difficult. Speaking as I-Mary, she suppresses the voice of Mary-by-herself when she describes her first encounter with the desert that was to become her home. In the third person voice of *Earth Horizon*, she writes: "All that long stretch between Salt Lake and Sacramento Pass, the realization of presence which the desert was ever after to have for her, grew upon her mind, not only the warm tingling presence of wooded hills and winding creeks, but something brooding and aloof, charged with dire indifference, of which she was never afraid for an instant" (182). I-Mary's voice in *Earth Horizon* can describe what might be frightening, but cannot admit to being afraid. Only privately could she recall these early experiences with the voice of Mary-by-herself. In "The Friend in the Wood" (intended to be a private autobiography of her relationship with nature), she describes her introduction to the desert and mountains while also admitting to her fear:

> There was one special time of its coming that stands as the type of all terror. That was after I had left the pleasant middle-western wood where I began, and had come to the desert outposts of the

> eastern slope of the high Sierras. There, astonishingly, with cold sickness fear had come to me in flashes. Walking or riding, out of utter desuetude, instantly the land would bare its teeth at me, crouch, tighten for the spring, and before the sweat of terror was dried, fall into desuetude again. (190)

Recalling her young womanhood in this essay, Austin allows the voice of Mary-by-herself to reveal her terrified reaction to a natural world she found threatening.

Less typically, this sense of fear intrudes on a public text in a description of the Sierra Madre mountains in *California: Land of the Sun*:

> [T]he land rested. And all in the falling of a leaf, in the scuttle of a horned toad, in the dust of the roadway, it lifted into eerie life. It bared its teeth; the veil of the mountain was rent. Nothing changed, nothing stirred or glimmered, but the land had spoken. As if it had taken a step forward, as if a hand were raised, the mountain [Sierra Madre] stood over us. And then it sank again. . . . Shall not the mother of the land do what she will with it? (39–40)

The lioness broods and suddenly turns, repeating the image of bared teeth. Significantly, this is the Madre (mother) mountain, unpredictable and unyielding, who in her beauty might nonetheless turn and bite. As articulated by Mary-by-herself, nature threatens violence on her human children. Intruding on an otherwise serene recollection of the beauties of the California landscape, the voice of Mary-by-herself reveals its still-potent fear of nature and its feelings of helplessness and passivity.[7]

With the voice of I-Mary, Austin consciously turned her back on a rational, science-based knowledge of nature in exchange for knowledge based on intuitive readings of the natural world. In *Earth Horizon*, Austin rejects what she called the "male ritual of rationalization in favor of a more direct intuitional attack" (15). This rejection reveals Austin's own feelings of connection to the natural world. In "The Friend in the Wood," she tries to explain how she was umbilically connected to the land:

> Even stick and stone, as well as bush and weed, are discovered to be charged with an intense secret life of their own; the sap courses, the stones vibrate with ion-shaking rhythms of energy, as I too shake inwardly to the reverberating tread of life and

> time.... Now and again I have that sharp impress of virtue in a plant or beauty in a flower, of which the Indians say, "It speaks to me... I begin to be aware of the soundless orchestration of activities beyond the ordinary reach of sense, the delicate unclasping of leaves of grass, the lapsing of dry petals into dust, the dust itself impregnate [sic] by the farthest star. (196)

Rather than being a mere objective observer of the natural world, I-Mary claims that she is part of this world, and that what she writes is shaped by her intimacy with it. Being subjective, for Austin, does not delegitimize her nature writing, as some critics have claimed, but instead serves as the very quality which gives her writing both authority and authenticity.[8]

How the land affects her, how she personally responds to the land: these, Austin claims, are the legitimate sources of her knowledge of the land about which she writes. Looking back on her career in her autobiography, Austin complains about how her approach has been misunderstood: "All the public expects of the experience of practicing Naturists is the appearance, the habits, the incidents of the wild; when the Naturist reports upon himself, it is mistaken for poeticizing" (*Earth Horizon* 188). She insists that she is not "poeticizing," that is, using nature to inspire her expression of emotion. This is a fairly radical departure from the literary status quo, as Austin began writing in the age of realism and naturalism, when nature writing began to be taken to task for its scientific inaccuracies.[9] Austin claims that she gives her readers an authentic version of the natural world because she uses the experiences of her body and emotions as the filter through which the natural world is distilled.

As I-Mary, Austin separates herself not only from the largely male tradition of nature writing, but also from most other women in her ability to connect with the natural world. In "The Friend in the Wood," she explains, "Women, I suspect, withdraw themselves from the appeal of the Wild because it distracts them from that compacted rounding of themselves which is indispensable to the feminine achievement" (187). Women, according to Austin, are frequently too caught up with the business of being an ornament to male culture to pay much attention to nature. In *Lost Borders*, she elaborates on those kind of women: "Women, unless they have very large and simple souls, need cover; clothes, you know, and furniture, social observances to screen them, conventions to get behind; life when it leaps upon them, large and naked, shocks them into disorder" (165). For Austin, Mary-by-herself belongs to this category.

Elsewhere, Austin has labeled such women bound by convention "ladies," finding them inferior to authentic women—women with "large and simple souls"—who live free from social restrictions and enjoy a powerful connection to nature. In an unpublished article entitled "If Women Did," she explains her belief that there exists for every woman, underneath her construction as a lady, a distant memory of her power as a woman. Austin asserts this power as a kind of primal source that connects women to the spirit of the earth:

> Among millions of modern women the cosmos still works, the *wokonda* [spirit] of the earth is a felt activity. But among other millions these influences have been cut off by the intervention of the falsely flattering persuasion of their ladyhood. Around the natural activities of women has been erected a series of reactions and inhibitions based upon the exemptions of their preciousness, of a commendable fragility and withdrawal.

Authentic women who reject the limitations and protections of "ladyhood" are not afraid to absorb the wild through their skin, to feel its rhythms. I-Mary is this kind of authentic woman, connected to the spirit of the earth through the experiences of her body and emotions.

With I-Mary, Austin abandoned her mother, rejecting her mother's choices and persisting in becoming a writer. In the *Tejon Notebook*, Austin puts her mother in the category of "ladyhood" but places herself in her category of natural woman: "I think the openness of everything scares mother. She has always lived in towns. And it frets her that I am not homesick. I can not make her understand that I am never homesick out of doors." Unlike her mother, I-Mary needs no shelter but the sky, nor any of the accessories of ladyhood in order to be at home there.

Models for I-Mary, women with "large and simple souls," can be found in the characters of Walking Woman in *Lost Borders* and Seyavi in *The Land of Little Rain*. More than fictional characters, these were women Austin knew in Kern County who became the basis for Austin's chisera figure—a wise woman, or prophet, who appears throughout Austin's work.[10] Austin describes this type of character in *The Flock* as someone who had made some "acknowledgment of the power of the Wild to effect a social divorcement without sensible dislocation" (261). Both Seyavi and Walking Woman had separated themselves from the Victorian world which confined women, but they were not lost; they were free to explore their desire.

The Walking Woman is free from the social constraints that normally bind women. Faith Jaycox writes that this figure is "the precise opposite of a confined woman: the endless mobility . . . is a powerful symbolic challenge to the enforced physical restriction of women at a moment in history when they had only recently worn clothes designed to suggest that they 'glided' rather than walked on two feet" (9). Austin explains how her character shed the conventions of ladyhood to become a woman: "She had walked off all sense of society-made values, and, knowing the best when the best came to her, was able to take it" (*Lost Borders* 261). The Walking Woman's freedom is not based on sexual neutrality; she has had the experiences of female sexuality and knows her body as a woman through love and motherhood, but has rejected the constraints of ladyhood. Yet the Walking Woman acquires her freedom only in the loss of her man and child, what David Wyatt calls an "erotic loss" that she replaces with desire; her walking marks her physical relationship with the female earth (90). She connects to the landscape by resituating her human desires within the geography inscribed by her footsteps. By walking, she has been "sobered and healed . . . by the large soundness of nature" (257). For Austin, the Walking Woman possesses a utopian relationship with nature, in which desire makes possible communion with the landscape.

Like the Walking Woman, Seyavi, the Paiute basket maker, is self-sufficient and has reduced her needs and wants without eliminating desire.[11] While the Walking Woman provides Austin with a model for the ideal intuitive relationship to nature, Seyavi takes Austin a step further and provides her with an intuitive form of creativity, of art, and for I-Mary, of writing. Seyavi's creativity comes not from the mind, but the body: "Every Indian woman is an artist,—sees, feels, creates, but does not philosophize about her processes. Seyavi's bowls are wonders of technical precision, inside and out, the palm finds no fault with them, but the subtlest appeal is in the sense that warns of humanness in the way the design spreads into the flare of the bowl" (*Land* 106). Seyavi transforms the impersonal geometry of design by weaving into her baskets her experiences of loss and love. Austin believes that rather than "philosophizing" about her creativity, Seyavi is connected to a tribal consciousness and her own unconscious, which invigorates her work—a work that moves according to the rhythms of her body and the rhythms of the desert landscape.[12]

Seyavi achieves what Austin desires for herself as an artist: to be so connected to the spirit of the land (the *wokonda*) that she can write through her body, not her mind. Only then will experience supersede analysis in her understanding and representation of the natural. Desire, Austin reveals, enables Seyavi's communion with the life of the land and her creativity. Authentic art must be created by body and emotion because, according to Austin, we create art in order to satisfy desire: "Seyavi made baskets for the satisfaction of desire,— for that is a house-bred theory of art that makes anything more of it" (*Land* 107). In Austin's "natural" theory of art (as opposed to "house-bred" theories), creativity is motivitated by the complex matrix of emotion and body that creates desire, at once erotic, primal, and innocent. I-Mary is modeled on Seyavi; her creative power, in which the body and emotion are more important than the intellect, is driven by desire.

Not surprisingly, the emotion that dominates I-Mary's voice is desire, the craving for fulfillment of body and soul through connection to the land. In *Lost Borders*, she admits: "Only Heaven, who made my Heart, knows why it should have become a pit, bottomless and insatiable" (174). For example, in *The Land of Little Rain*, I-Mary's voice admits to desire, to needing more, to longing to satisfy her "keen hunger," to wanting to know her way into the secret heart of the unresponsive desert (108). This becomes clearest in the passage about her desire to learn the "secrets of plant powers": "I remember very well when I first came upon a wet meadow of *yerba mansa*, not knowing its name or use. It *looked* potent; the cool, shiny leaves, the succulent pink stems and fruity bloom. A little touch, a hint, a word, and I should have known what use to put them to. So I felt, unwilling to leave it until we had come to an understanding" (144).[13] For Austin, such knowledge is not acquired through study but through her desire for "an understanding," for some sort of intuitive connection.

I-Mary's desire for connection appears in various depictions of nature, some of which make clear that this nature is female. At times, such metaphorical transformations border on the homoerotic, as in this passage from *California: Land of the Sun* (1914):

> A pomegranate is the one thing that makes me understand what a pretty woman is to some men. . . . The flower of the pomegranate has the crumpled scarlet of lips that find their excuse in simply

> being scarlet and folded like the petals of a flower; and then the fruit, warm from the sunny wall, faintly odorous, dusky flushed! It is so tempting when broken open . . . the rich heart colour, and the pleasant, uncloying, sweet, sub-acid taste. (90)

Here, I-Mary evokes the seductive power of women in a highly eroticized image of female genitalia and claims to understand women's power of seduction as a man would, with physical hunger for a taste of beauty.[14] Rarely does I-Mary's desire take on such an erotic form. More frequent in Austin's writing are passages which blend mildly sensual images of the landscape with a sense of loss. Speaking of the "musky and sweet" gilia blooming on her path in her unpublished story, "The Lost Garden," she writes: "Some scents there are that come to us so freighted with a sense of remembered beauty as sets the mind agrope for a warrant for it . . . sweet with the pressage [sic] of satisfaction" (2).[15] In this image, beauty and desire come together in nostalgia and longing. John O'Grady explains: "Desire in Austin's work is always predicated upon loss—lost borders, lost loves, lost lives" (129). At the same time, the gilia represents not simply a yearning for remembered beauty, but a living beauty that foretells the coming fulfillment of desire. In these eroticized images, I-Mary's desire for a connection with a feminized nature is made possible.

I-Mary also transforms Mary-by-herself's lioness from a representation of nature as frightening into a figure of desire, like the pomegranate and gilia. This transformation suggests I-Mary's desire not only for connection to the feminine in nature, but also for completion of herself as a woman: mythical, powerful, fulfilled. Perhaps the best known rendition of the lioness appears merged with an image of Austin's idealized self early on in *Lost Borders*:

> If the desert were a woman, I know well what she would be like: deep-breasted, broad in the hips, tawny, with tawny hair, great masses of it lying smooth along her perfect curves, full lipped like a sphynx, but not heavy-lidded like one, eyes sane and steady as the polished jewel of her skies, such a countenance as should make men serve without desiring, such a largeness to her mind as should make their sins of no account, passionate, but not necessitous, patient—and you could not move her, no, not if you had all the earth to give, so much as one tawny hair's-breadth beyond her own desires. (160)

As photos of Austin reveal, this goddess even looks like Austin, sharing her thick tawny hair and full lips. Now a sphynx, this lioness represents female strength, passion, and self-assuredness. Both seductive and indifferent, this goddess is powerfully compelling.

In relation to men, I-Mary's lioness is far different than the one that confronts Mary-by-herself. Gone are the images of bared teeth; instead of an aggressor, the sphynx-lioness is seductive but not a seductress; she is basically indifferent to the male desire to possess her. This version of the lioness powerfully toys with the men who want to control her and finally rejects them. She seduces the male sojourning in the desert; in the short story, "The Hoodoo of the Minnietta," a man is described as caught in her power: "[T]he desert had him, cat-like, between her paws" (*Lost* 164). In *The Flock*, she awaits the emotional disintegration of the man faced with her indifference: "But over the faces of the men whose life is out of doors . . . comes the curious expression which is chiefly the want of all expressiveness. . . . It is as if one saw the tawny land above them couched, lion-natured, lapping, lapping" (256). Men clearly have no control of this creature. In her novella, *Cactus Thorn*, she explains the hypnotic power of the desert wild over men: "[O]ne perceives the lure of the desert to be the secret lure of fire, to which in rare moments men have given themselves as to a goddess" (8). This goddess is the sphynx-lioness who embodies desire.

With the lioness, Austin not only represents her own fears and desires, but she also uses this metaphor to critique the traditional (male) relationship to nature as a power struggle through which men try to control nature. Austin observes that rather than a love of the land, men have "the love of mastery, which for the most part moves men into new lands" (*Lost* 192). For Austin, this is not only the "mastery" of dams and fences, but also the scientific approach to the natural world that tries to diminish its power by dissecting it intellectually. For men, the desire that the land stirs is for possession and control; the result, according to Austin, is eternal damnation, for the men who face the lioness are destroyed by their unsatisfied longing. For authentic women, the desire that the land stirs is for connection. Austin's ideal women come to know the land through intuition and the experiences of their bodies—think of the Walking Woman's footsteps, Seyavi's baskets, or the written image of I-Mary's pomegranate. Their desire is fulfilled not through an actual embrace of the female body by the land, but through the creative processes this feminized nature stimulates in women.[16] When Austin merges the sphynx with I-Mary, she is merging the voices of desire and creativity, and placing both out of men's grasp.

Austin's desire for a connection to this feminized land is her desire for both the creative fulfillment of self as well as a connection with the maternal, which reenacts her relationship with her mother—whose approval she could never win, whose affection seemed so tentative. In her autobiography, Austin explains that she was an unwanted child, who "was not desired, not, in fact, welcome" (*Earth Horizon* 32). Austin's experience of her mother was one of longing: a hunger for a portion of herself, a reflection of herself in her mother's love and reassurance. Austin depicts her mother as a woman who demanded her free-spirited daughter's conformity to the restrictions of Victorian femininity and judged her harshly when she failed to meet this standard. This experience created Mary-by-herself.

For Mary-by-herself, union with the maternal is a ferocious devouring of the child by the mother. I-Mary replaces longing for the union with the mother with a desire for an umbilical connection with a female-gendered nature that will animate her own creative processes. Elizabeth Ammons observes: "As Austin's language suggests . . . her need for art to be physical, to hold the body as well as the spirit, contains powerful erotic longing. Particularly, her need suggests the ferocious hunger of the infant, in this case the daughter, for the unconditional and intensely physical love of the mother" (90). Austin seems to have experienced her own mother as either a brooding presence that motivated her fear, or an absence that stimulated her desire. Austin herself recognized that desire is stimulated by absence and that hunger for connection creates art. Commenting in *Everyman's Genius* on how many artists are unwanted children, Austin writes, "this craving for communication, this tormenting desire to tell, is merely the emotional register of the inward drive which is indispensible to all creative success" (111). Therefore, I-Mary's creativity and desire for connection depend on Mary-by-herself's longing and fear of abandonment.

We can see that Austin's efforts to write nature emerge from a complex matrix, and her desire to write is complicated by the tension between being the woman she wanted to be and bearing the burden of ladyhood that her mother had passed on to her. Divided by the fetters of convention and her own desires for communion with the female embrace of the land, her narrative voice is fueled by the tension this division creates. Austin takes as the touchstone of her work her own experience of the desert and mountains. In this landscape so indifferent to human concerns, Austin replaces the tropes of the nymph and earth mother with the lioness, transforming the notion of nature's female power into something fierce and passionate. Austin's dual response to these qualities marks her narrative voice, refiguring the desert and mountains of the Southwest as well as the shape of the natural history essay. Austin's writing looks forward to the subjective visions found in the work of Annie Dillard, Barry Lopez, and others, who also narrate their relationships to the natural world through a web of unnamed private experience and history.

Notes

1. See Ammons, Fink, Graulich, Langlois, Lanigan, Morrow, and Pryse.
2. Cheryl Walker has identified a useful term for the critical method I will employ: "persona criticism," which she defines as "a critical practice that both expands and limits the role of the author . . . finding in the text an author-persona but relating this functionary to psychological, historical, and literary intersections quite beyond the scope of any scriptor's intentions, either conscious or unconscious. The persona functions more like a form of sensibility in the text than a directional marker pointing back toward some monolithic authorial presence" (114).
3. See Wyatt in particular. Also, Karen Langlois's article on Austin shares this point of view.
4. For a more complete discussion of Austin's relationship with her mother, Susannah Hunter, see Lanigan Stineman's biography of Austin.
5. Other critics of Austin have observed this figurative strategy as well; Shelley Armitage writes: "Seeking to awaken this society to truths visible in its own landscape, yet ignored, commercialized, or managed without stewardship for profits, she often characterizes the land as feminine, applying the Native American concept of Mother Earth. . . . The land is nurturing, spiritual, and resilient" (21).
6. Mary Hunter married Stafford Wallace Austin in 1891 and soon found her husband's inability to make a living a burden. They were often separated, even after the birth of their daughter, Ruth, in 1892. Austin was frequently left alone to support herself and her disabled

daughter, which she did through teaching and, eventually, writing. She left her husband permanently in 1904 (although they did not divorce until 1914) and made the difficult decision to institutionalize her daughter a year later.

7. I can only speculate what brought about the return of Mary-by-herself in a piece of Austin's published work, as Austin had only permitted this voice to be exposed in letters and poems. After the publication of *The Land of Little Rain*, Mary-by-herself is often heard in Austin's private papers when Austin discusses illness, professional discouragement, and problems with relationships. It is interesting to note that Austin finally obtained a divorce from her husband in 1914, the same year *California: Land of the Sun* was published.

8. See Blend and Scheick.

9. In 1895, naturalist John Burrough's wrote "the literary naturalist does not take liberties with facts; facts are the flora upon which he lives. . . . To interpret Nature is not to improve on her: it is to draw her out" (quoted in Lutts 9).

10. See Graulich.

11. Vera Norwood comments that "Seyavi provides Austin with a human incarnation of the spirit of the feminine southwest land and since for Austin, under the best circumstances, land culture and human culture are one, the best human method for living on that land" (14).

12. See especially *The American Rhythm* (1923) for Austin's theories about Native American creative processes and the tribal consciousness which enables their art.

13. Austin is referring to *la yerba del manso*, a Hispanic folk cure for burns, sores, colic, and dysentary.

14. Paula Bennett identifies what has been called "the Language of Flowers" as a language "through which women's body and . . . women's genitals have been represented and inscribed" (240).

15. Austin writes on the cover page to this manuscript: "This is the original sketch rewritten in 1906 from note[s] made in 1902. In 1912 it was again rewritten and slightly revised in 1928 for Albert Bender."

16. In a very different reading, Ammons believes that Austin's search to recover the mother—"the Mother Earth, the mother artist, the mother tongue of the land" is "gentle and earthbound," and "her experimental prose rocks and caresses" the reader, rather than challenges the reader, as I believe Austin's work does (97).

Works Cited

AMMONS, ELIZABETH. *Conflicting Stories: American Women Writers at the Turn into the Twentieth Century*. New York: Oxford UP, 1991.

ARMITAGE, SHELLY, ED. "Writing Nature." Introduction to *Wind's Trail: The Early Life of Mary Austin*. Santa Fe: Museum of New Mexico P, 1990.

AUSTIN, MARY HUNTER. *The American Rhythm*. 1923. New York: AMS, 1970.

———. *Cactus Thorn*. Ed. Melody Graulich. Reno: U of Nevada P, 1988.

———. *California: Land of the Sun*. New York: Macmillan, 1914.

———. *Earth Horizon: An Autobiography*. 1932. Albuquerque: U of New Mexico P, 1991.

———. *The Flock*. Boston: Houghton Mifflin, 1906.

———. *Everyman's Genius*. Indianapolis: Bobbs-Merrill, 1923.

———. "The Friend in the Wood." In *Wind's Trail: The Early Life of Mary Austin*. Ed. Shelley Armitage. Santa Fe: Museum of New Mexico P, 1990: 183–98.

———. "If Women Did" [1918<>1934]. Ts. AU 232. Austin papers. Huntington Library, San Marino, CA. N. pag.

———. *Inyo* [1899]. Ts. AU 261. Austin papers. Huntington Library, San Marino, CA.

———. *The Land of Little Rain*. 1903. Albuquerque: U of New Mexico P, 1974.

———. *Lost Borders. Stories from the Country of the Lost Borders*. Ed. Marjorie Pryse. New Brunswick, NJ: Rutgers UP, 1987. 151–263.

———. "The Lost Garden" [1906]. Ts. AU 318. Austin papers. Huntington Library, San Marino, CA.

———. *Tejon Notebook* [1889?]. Ms. AU 267. Austin papers. Huntington Library, San Marino, CA. N. pag.

BENNETT, PAULA. "Critical Clitoradectomy." *Signs* 18.2 (Winter 1993): 1–27.

BLEND, BENAY. "Women Writers and the Desert: Mary Austin, Iza Sizer Cassidy, and Alice Corbin." Ph.D. diss., U of New Mexico, 1988.

FINK, AUGUSTA. *I-Mary: A Biography of Mary Austin*. Tucson: U of Arizona P, 1983.

FLAX, JANE. "Mother-Daughter Relationships: Psychodynamics, Politics, and Philosophy." *The Future of Difference*. Ed. Hester

Eisenstein and Alice Jardine. 1980. New Brunswick, NJ: Rutgers UP, 1988: 20–40.

GRAULICH, MELODY. Afterword. *Earth Horizon: An Autobiography*. By Mary Austin. Albuquerque: U of New Mexico P, 1991.

JAYCOX, FAITH. "Regeneration through Liberation: Mary Austin's 'The Walking Woman' and Western Narrative Formula." *Legacy: A Journal of Nineteenth-Century American Women Writers* 6.1 (Spring 1989): 5–12.

LANGLOIS, KAREN. "Mary Austin's *A Woman of Genius*: The Text, the Novel, and the Problem of Male Publishers and Critics and Female Authors." *Journal of American Culture* 15.2 (Summer 1992): 79–86.

LUTTS, RALPH H. *The Nature Fakers: Wildlife, Science, and Sentiment*. Golden, CO: Fulcrum, 1990.

MORROW, NANCY. "The Artist as Heroine and Anti-Heroine in Mary Austin's *A Woman of Genius* and Anne Douglas Sedgewick's *Tante*." *American Literary Realism, 1870–1910* 22.2 (Winter 1990): 17–29.

NORWOOD, VERA. "The Photographer and the Naturalist: Laura Gilpin and Mary Austin in the Southwest." *Working Paper #6*. Tuscon: Southwest Institute for Research on Women, 1981.

O'GRADY, JOHN P. *Pilgrims to the Wild: Everett Ruess, Henry David Thoreau, John Muir, Clarence King, Mary Austin*. Salt Lake City: U of Utah P, 1993.

FRYSE, MARJORIE. Introduction. *Stories from the Country of Lost Borders*. By Mary Austin. New Brunswick, NJ: Rutgers UP, 1987: vii–xxxviii.

SCHEICK, WILLIAM J. "Mary Austin's Disfigurement of the Southwest in *The Land of Little Rain*." *Western American Literature* 27.1 (Spring 1992): 37–46.

STINEMAN, ESTHER LANIGAN. *Mary Austin: Song of a Maverick*. New Haven: Yale UP, 1989.

WALKER, CHERYL. "Persona Criticism and the Death of the Author." *Contesting the Subject: Essays in the Postmodern Theory and Practice of Biography and Biographical Criticism*. Ed. William Epstein. West Layfayette, IN: Purdue UP, 1991: 113–19.

WYATT, DAVID. "Mary Austin: Nature and Nurturance." *The Fall into Eden: Landscape and Imagination in California*. Cambridge: Cambridge UP, 1986: 68–94.

Misogyny in the American Eden
Abbey, Cather, and Maclean

J. GERARD DOLLAR

> Why, we ask ourselves, floating onward in effortless peace deeper into Eden, why not go on like this forever? True, there are no women here (a blessing in disguise?) . . .
>
> EDWARD ABBEY, *DESERT SOLITAIRE* (183)[1]

Edward Abbey tries to distance himself, at least a little bit, from this misogynistic comment in the "Down the River" chapter of *Desert Solitaire*. He frames his thought as a question rather than a statement, and he places it in parentheses. Just another of Abbey's outrageous and provocative asides, the reader assumes. But maybe not.

Readers of Abbey's 1968 classic will recall that "Down the River" describes Abbey's magical and Edenic journey through the doomed Glen Canyon of southern Utah. Abbey and his good friend Newcomb float lazily, dreamily, sometimes pensively down the Colorado River beneath the towering walls of the magnificent canyon. They swim naked in the river, snare catfish for their dinners, explore Anasazi cliff dwellings; and they experience profound peace as they wonder at the great, silent beauty of this unknown wilderness. But is the escape from women, which this journey affords, really tangential to the paradisal quality of the experience, as Abbey's parentheses seem to imply? Or does Abbey reflect a tendency in American nature writing, especially writing about the American West, to define women out of paradise? Is it Eden, even though women are not there, or is it Eden *because* women are not there?

In considering briefly three twentieth-century versions of the western Eden—or, the wilderness Edenic—one finds that the presentation, and indeed veneration, of the western wilderness as the site for men both to escape women and to bond with other men—a mythos that we might ordinarily associate with Hemingway—is disturbingly widespread. It can be found in a wide range of texts, and not just in texts authored by men.

I begin with Abbey because in many ways "Down the River" represents a *locus classicus* in American nature writing. Here we find the western wilderness as paradise, the natural world as a place of refuge, an escape from a debased civilization and the site of self-discovery, even epiphany. Glen Canyon is Abbey's cathedral; this is where he comes face to face with the sacred. The idyllic river journey leads Abbey and his friend back to Eden, but it is Eden without Eve, an Eden of two Adams who are obviously delighted that the women have been left behind in the modern world of Albuquerque—clearly where they belong.[2] Women are very much part of that "other world" out there, from which these men are so eager to escape. The liberating wilderness experience Abbey ecstatically celebrates is founded on a wilderness/civilization (or purity/debasement) binary that he frequently invokes; but here, as well as in texts by Willa Cather and Norman Maclean, the male/female binary forms an insistent subtext. Explicit in these texts is a linking of the wilderness Eden to male experience and performance; implicit is an antithetical relationship between wilderness and women. The western Eden is valued to the extent that women are excluded and devalued.

In "Down the River" the sacred place in nature is a male space, desirable not just for its own beauty, or its potential for adventure and self-discovery, but because the women are left behind. Without women Abbey and Newcomb can become boys again; escaping women means a holiday from commitment, responsibility, family. Indeed, Abbey is quite self-conscious about the adolescent nature of his fantasy-voyage: "I am fulfilling at last a dream of childhood and one as powerful as the erotic dreams of adolescence—*floating down the river*. Mark Twain, Major Powell, every man that has ever put forth on flowing water knows what I mean" (176). Presumably every woman who has "ever put forth on flowing water" does not know what Abbey means. Abbey's is indeed a variation on Huck Finn's adolescent—and perhaps homoerotic—fantasy: to embrace nature, to love another man within the pure context of nature, and to escape the Miss Watsons of the world—the women who would civilize and shackle male freedom.

In this version of Eden freedom from all constraints obviously plays a large role. Abbey may define himself positively in terms of the natural world, but he also defines himself negatively—against the industrial, the modern, and the feminine. In the absence of women we even find Abbey feminizing Newcomb, much as Huck feminizes Jim. Newcomb is the cook; his place, like Jim's, is back at the raft, while Abbey wanders off on a series of dangerous exploratory adventures. Abbey may at times sound Wordsworthian in espousing a wise passiveness before nature, but his impulse is generally exploratory and performative. And male performance must here be validated by a feminized, but not female, "audience." Thus Newcomb is there waiting for him at the end of a long day, offering support, good companionship, and the amenities of a domestically minded partner with none of the emotional baggage that comes, Abbey implies, with male-female relationships.

The paradox in Abbey, and in my two other twentieth-century examples of the wilderness Edenic, is that the highest spiritual development, through contact with nature, is found alongside what Freud would surely label arrested development. The male's affirmation of a wilderness self comes at the price of denying or repressing a sexual self and a social self; it is as if the natural world becomes the man's true spiritual mate—an idealized womanly Other who makes flesh-and-blood women at best an irrelevancy, at worst a temptation away from "pure" male self-fulfillment.[3] Women as alternative mates—the temptress Eves who must be excluded from Eden—threaten the loss of self-transcendence. As in Genesis, they are the path out of Eden. Embedded in Abbey's "Down the River" is the idea that wilderness is not to be feared—but women are.

Consider next a work of fiction from early in the century. Willa Cather's *The Professor's House* (1925) offers another good example of the quest for a sacred space in nature. Here the quest is carried out primarily by Tom Outland, Cather's idealistic young cowhand who grows up in the Southwest and then travels east for a college education, an engagement to the beautiful older daughter of his beloved Professor St. Peter, and brief fame as a scientist before his early death in World War I. Readers will recall that "Tom Outland's Story," the middle section of the novel, is the window that opens up—following an account of the professor's claustrophobic and disillusioned life in a small college town—on a world of space, clarity, light, and freedom (all that the middle-aged professor feels he has lost). Tom's discovery of the beautiful Blue Mesa in northern New Mexico presents another version of the discovery of an Eden in the American West. Here we are in the realm not only of great natural beauty, but also of tranquility, peace, purity. Tom first has his vision of the cliff city through the purifying medium of a snowstorm on Christmas Eve, and there is a strong sense of the sacred, of epiphany, in what he sees.

In Cather's text, as in Abbey's, we see the extent to which women are excluded from Eden. Again, the attainment of this paradise is presented as a test of courage and resolve; Tom Outland must cross a dangerous river and then ascend a precipitous trail. (Water imagery in both Cather's and Abbey's works suggests a baptism into a new life.) The journey is fraught with dangers, leading to a sense that wilderness is a man's world. Tom is joined by his male companions, Roddy Blake and the elderly Henry Atkins, in exploring the mesa. The sense of danger—of being in a previously unknown world, of being pioneers—gives the men an intensified sense of camaraderie. When Tom later returns to the mesa, he brings with him both of his father surrogates: the Catholic priest, Father Duchene, and Professor St. Peter. There is never any mention of his wishing to show his fiancée, Rosamund, the Blue Mesa; apparently he keeps his two lives, and two selves, separate.[4] The original self, and pure self—as the Professor comes to realize in the novel's final chapters—predates and exists completely apart from the formation of any sexual relationships; the acknowledgment of one's sexual self represents a fall, a distortion of the original self.

The Professor's House contains at least one blatantly misogynistic aside, and it comes in the section detailing the Professor's growing bitterness over his wife's and daughter's materialism and acquisitiveness. "I was thinking," he tells his wife Lillian, "about Euripides; how, when he was an old man, he went and lived in a cave by the sea, and it was thought queer, at the time. It seems that houses had become insupportable to him. I wonder whether it was because he had observed women so closely all his life" (156). This same spirit carries over into the Tom Outland section of the novel. And it appears most clearly when Tom and his cohorts discover the mummy of the murdered Anasazi woman, which they name Mother Eve. This is the only female presence on the Blue Mesa, and she remains something of an enigma. It is certainly telling that Tom's friend Father Duchene is brought in to interpret and thereby "place" Mother Eve. The priest's hypothesis, to account for the ancient murder, is that "[p]erhaps her husband thought it worthwhile to return unannounced from the farms some night, and found her in improper company" (223). So from the perspective of this Catholic priest—a perspective obviously shared by both Tom Outland and the Professor—a woman's presence in paradise brings with it sexuality, temptation, and the risk of falling from pure male identity and male purpose.[5] Apropos of this theme, we recall that Tom's reading on the Mesa, when he spends his solitary summer there after the falling out with Blake, is Virgil's *Aeneid*—the epic poem which describes (among many other things) Dido's temptation of Aeneas and his resolute adherence to his male quest.

Cather was not only a devoted admirer of the classics but a great lover of opera as well, and one of her favorite works was Wagner's last opera, *Parsifal*. The Tom Outland story is in fact very reminiscent of Parsifal's quest for the Holy Grail—for the spiritual purity it represents—and in Wagner's opera the grail is guarded by a spiritual brotherhood. The only woman in the opera, Kundry, represents sexual temptation. Cather similarly has her Blue Mesa discovered and protected by a male brotherhood—Tom, the Father, the Professor, and Roddy Blake (until he proves unworthy of the sacred trust). There is no place for women on the Blue Mesa; in a novel which casts women in the role of diminishing male freedom, they are excluded from the visionary heights which Tom briefly attains before his fall into a sexual attachment to Rosamund St. Peter, an attachment which he might very well be fleeing when he goes off to fight, and die, in the Great War.

A much more recent example of the wilderness Edenic appears in Norman Maclean's *A River Runs Through It* (1976), certainly the most Hemingwayesque of these three texts. Here too we find the American Eden presented as a western wilderness landscape: a world of challenges where men go to test themselves; a world that separates strong men from weak men; a hard-to-reach place (just getting there is a test) and again a sacred realm where women don't belong. There is certainly much to discuss in Maclean's finely crafted novella, but consider one important scene that summarizes many of the work's themes. Norman, the book's first-person narrator, and his beloved brother Paul have had a fishing trip spoiled by the presence of

Norman's citified brother-in-law Neal—an egocentric, self-promoting character, who clearly does not belong in the rugged natural world of Montana fishermen. To purify themselves after this failed outing, Norman and Paul go off to "their" river, their sacred place in nature, the Big Blackfoot River, only to experience an even worse sense of desecration as Neal again decides to join them, this time bringing the town prostitute, Old Rawhide, along with him. After again testing their skills as fly fishermen, reading the river to determine which flies will draw the fish—and thus proving themselves both real fishermen and real men—Norman and Paul make two horrific discoveries: that their beer has been stolen and that Neal and his lady friend are lying naked, backsides severely burned, on a sandbar in the middle of their river. Now if the Big Blackfoot River represents an Eden, then Neal and his lady-friend are surely not out of place: naked lovers, lost in post-coital oblivion, totally free of all inhibitions or social constraints.

Obviously this is not the way that Norman and Paul see it, and what is striking in the scenes depicting their outrage is the extent to which they blame the woman—surely another example, comparable to Father Duchene's self-satisfied reading of Mother Eve, of blaming the victim. Neal, loathsome as he is to Norman, must be looked after as family; but Old Rawhide is the one who must be literally, violently kicked out of their lives. When Norman and Paul come upon the naked lovers they discover that Old Rawhide has "LOVE" tattooed on her backside—"LO" on one cheek and "VE" on the other. This becomes Paul's target when the brothers drop off Old Rawhide in town: "She was running barefoot and trying to hang on to the rest of her clothes and [Neal's] underwear, so Paul caught up to her in about ten jumps. On the run he kicked her, I think, right where the "LO" and the "VE" came together. For several seconds both of her feet trailed behind her in the air. It was to become a frozen moment of memory" (79). So this ritual expulsion of the woman by Paul becomes another of Norman's precious "frozen moments"—moments that invariably present the beauty of Paul, moments that usually show him fishing expertly, phallic rod in hand. Of course Paul has a succession of women friends, but the implication is that one's relations with women belong to another, lesser world than the world of pure, spiritual fulfillment and love between men.

Old Rawhide's troubles don't end with Paul, either. Norman is so moved by the moral and symbolic rightness of Paul's actions that he must try to follow suit: "Suddenly, I developed a passion to kick a woman in the ass. I was never aware of such a passion before, but now it overcame me. I jumped out of the car, and caught up to her, but she had been kicked in the ass before and by experts, so I missed her completely. Still, I felt better for the effort" (80). All of the anger, violence, and smug sense of superiority which Norman directs against Old Rawhide comes within the context of (and is no doubt accounted for by) his considerable fear—and awe in the presence—of his wife and mother-in-law. Several passages in the novella mythify the strength, resolve, and altogether godlike qualities of the Scottish woman; and never did Huck Finn or Tom Sawyer quake before the stern discipline of Aunt Sally or Aunt Polly as much as Norman trembles before the anticipated wrath of his wife and mother-in-law. These passages begin with Norman's reverence for his mother, and in the work's early scenes we note that the purity of the mother comes in part from her knowing her place—that is, the house. She does not venture into the sacred realm of the men's trout river, another males-only Eden. The debased heterosexual love which Norman and Paul seem to associate with a woman's backside must be driven from their paradise; that profane word "LOVE" must be booted out so that the river can again speak clearly to them its sacred words.

All of these texts therefore present Edenic versions of the American West from which women are clearly and emphatically excluded. Rather than Adam and Eve in these pristine surroundings we instead find Adam and Adam: Abbey and Newcomb, Tom Outland and Roddy Blake, Norman and Paul. Women are not wanted here, and if they intrude they are viewed with mistrust. All three works show tremendous yearning and nostalgia for a pure source in nature—an untainted original world; all rely heavily upon water imagery to suggest purity, cleansing, renewal; and all three show love between men—with a tendency within the narrative voice to depict the other man as weaker, closer to the fallen world of women. Interestingly, all three texts extend the quest for origins to literature as well as nature: Tom Outland atop the Blue Mesa pores over his sacred text, *The Aenead*, while Abbey evokes Twain and Powell. Each text represents a version of the male journey to find himself and prove himself. In *A River Runs Through It* the sacred text is obviously the Bible, with most of the emphasis falling on the Old Testament, Genesis in particular. The Big Blackfoot cuts through to the oldest stones in the world, to an avatar of the beginning, and in the beginning was the word; the true initiate, like Norman, can hear the river speaking with the voice of God (the father), speaking the word. If indeed the word is "love," as the novella suggests, there are clearly two kinds of love here: the "pure" brother-brother and father-son love of the Maclean men, and the debased "LOVE" inscribed on Old Rawhide's raw hide and exposed to the outraged brothers. This Eden obviously has no place for this particular Eve—or any Eve, one suspects.

Ecofeminists and others have rightly critiqued an American "master narrative" in which active, privileged male will subjugates and conquers passive female nature.[6] What is interesting in these texts is that all three male narrators—Abbey, Tom Outland, and Norman—appear to be dissenting from this rapacious and male-centered mythos. All three are preaching respect for the land and voicing a modern wilderness ethos of taking only memories from the wilderness Eden and leaving only footprints. Yet the male will-to-power is present, albeit in a more covert form. For these explorers and questers after Edenic peace are also writers, and in their texts they have possessively inscribed their Edens as male space, their space. The borders that mark off Eden from the debased world below it or outside it also serve to keep out those whose presence might desecrate these holy realms—and foremost among these threatening intruders are women.

Notes

1. The sentence I have truncated includes a long list of what Abbey and Newcomb, in rafting down Glen Canyon, will (happily) have to do without; the list includes concert halls, books, bars, theaters, wars, and traffic jams. Abbey thus tries to cover both the "blessings" and banes of civilization, with the status of women here left somewhat ambiguous (although there is certainly an element of damning by association).
2. In an ironic and self-congratulatory vein, Abbey asks Newcomb, "do you think it's fitting that you and I should be here in the wilds . . . while our wives and loved ones lounge at their ease back in Albuquerque, enjoying the multifold comforts, benefits and luxuries of modern contemporary twentieth century American urban civilization?" (181–82).
3. In her introduction to *The Lay of the Land*, Annette Kolodny describes "what is probably America's oldest and most cherished

fantasy: a daily reality of harmony between man and nature based on an experience of the land as essentially feminine—that is, not simply the land as mother, but the land as woman, the total female principle of gratification—enclosing the individual in an environment of receptivity, repose, and painless and integral satisfaction" (4).

4. Tom Outland and the Professor are similar in many ways—they could be father and son—and we note that the older man also has an Edenic place in nature, Lake Michigan, where he can lose himself in order to find himself; similarly, we never see the professor at the lake with his wife or daughters.

5. Note that Mother Eve, who parallels the original Eve in suggesting (to men) women's ability to make men fall, literally becomes a fallen woman when the donkey transporting her off the mesa falls down into the canyon below (244); she had been sold by Roddy Blake to a German collector of artifacts, but here the male urge to possess the spirit of the mesa and erase women from this sacred place meets with defeat. The spirit of Mother Eve will remain on the mesa, suggesting the potential for a return of the repressed.

6. See, among others, Kolodny, Ruether, and Merchant.

Works Cited

ABBEY, EDWARD. *Desert Solitaire*. New York: Ballantine, 1968.

CATHER, WILLA. *The Professor's House*. 1925. New York: Vintage, 1973.

KOLODNY, ANNETTE. *The Lay of the Land: Metaphor as Experience and History in American Life and Letters*. Chapel Hill: U of North Carolina P, 1975.

MACLEAN, NORMAN. *A River Runs Through It and Other Stories*. 1976. New York: Pocket, 1992.

MERCHANT, CAROLYN. *The Death of Nature: Women, Ecology, and the Scientific Revolution*. San Francisco: Harper & Row, 1980.

RUETHER, ROSEMARY RADFORD. *New Woman/New Earth: Sexist Ideologies and Human Liberation*. New York: Seabury, 1975.

The Body as Bioregion
DEBORAH SLICER

So far, ecological feminists and bioregionalists, two of the several representatives of the radical environmental movement, have had a congenial but rather shallow relationship grounded in a celebration of the domestic. Judith Plant, an ecofeminist and bioregionalist, commented on that relationship recently in *Home! A Bioregional Reader*. She suggests there that women, whose traditional purview has been the domestic, must teach men the value of home and teach them how to put our social and ecological households in order. Ecofeminist philosopher Val Plumwood calls these women, sardonically, "angels in the ecosystem" (10). To many a weary housemarm this role sounds all too familiar. And to those women who have spent years honing their survival skills, Plant's recommendation may even sound dangerous. After all, in the world as it is women are neither safe nor in control in the household, as even Plant notes. In fact, physical violence against women occurs most often in women's homes, where family members are most likely to view and physically lay claim to women as domestic property.

We will probably never know which was the historical first and paradigm for the other: the violation of a woman's body or of the land. In a culture that defines both the human female body and the land as "resource," as someone else's "property," such violations are conceptually guarded secrets. Sowing oats in a "barren" field or in a woman is to put something otherwise wasted into production. Social meanings can physically coerce the body and most certainly influence our experiences of the body. But sometimes the land, animals, or women—supposedly inert matter—refuse these meanings, crossing over into a conceptual wilderness which might entail prosecuting a rapist or yielding knapweed instead of oats. When that happens we find ways of coercing them with more familiar means and meanings, with a heavy hit of Tordon, a carcinogenic herbicide, for example, or we send women through a legal system that cannot distinguish between sexual passion and assault.

My sense is that before it's safe for either women or men to go back into the home, even in the broader, environmentalists' sense of home as one's most proximate ecological bioregion, we must come to terms with the complex and destructive social meanings of the body, of that ecosystem with which we are self-identical and about which most of us are virtually ignorant. And when those social meanings fail us, and most of them will because they are mapped onto other meanings that environmentalists are regularly critical of, we will reclaim something of ourselves, or at least be poised to, as we enter that great wilderness of our own corporeal being.

❀ ❀ ❀

In 1976, poet and essayist Adrienne Rich said that "I know no woman—virgin, mother, lesbian, married, celibate—whether she earns her keep as a housewife, cocktail waitress, or a scanner of brain waves—for whom her body is not a fundamental problem: its clouded meaning, its fertility, its desire, its so-called frigidity, its bloody speech, its silences, its changes and mutilations, its rapes and ripenings" (284). Corporeal life in general is an embarrassment to a rationalist metaphysics. And the female, which is often equated with corporeal life, is only one of many bothersome reminders of corporeal existence and of the duplicitous illusions of Plato's cave. My mother's body is my most fundamental, problematic heritage, to paraphrase Rich (284).

It's a strange feeling to not only have outlived my mother but to have lived longer than my mother. My mother died when she was thirty-seven and I was twelve. I am now thirty-nine, two years older than she was when she died of cancer that began in her breast and eventually spread to her bone and then to her brain. When, two years after radiation, the disease reappeared in her bones and they began to break, she was told she had arthritis. My father knew the truth. He consulted with the doctors about her treatments and he, not my mother, decided when it was time to stop the treatments, to allow her to die. I understand that this sort of paternalism was quite common twenty-five years ago. My mother must have known that what was killing her wasn't arthritis, but I never heard the word "cancer" pass her lips. She almost never let me help her, not even to the bathroom. I was the child and she was the mother. She would insist on at least this pretence of normalcy in her life and on this small arena of control. There is no unambiguous good or evil in this story. My mother was not a helpless victim. She asserted her will in the few ways that were available to her, ways that I am only now understanding.

For all these years research on breast cancer has been funded at a small pittance of other less common, less deadly cancers. And the research that is funded focuses on technological and chemical fixes rather than prevention, diet, or environmental hazards, where, should we find a cause and thus a cure, there would be no long-term future for pharmaceutical companies and others in the research industry, who are grossing over $100 billion a year dabbling in cancers (Clorfene-Casten 57). Not coincidentally, the American Cancer Society, which funds a good portion of this research, has failed to support such environmental legislation as the Clean Air Act, the Toxic Substances Control Act, clean water legislation, and efforts to reduce radiation exposure (Clorfene-Casten 57). The National Cancer Institute has a similar environmental record, and the President's Cancer Panel, which advises Congress, was chaired for the last decade by the late Armand Hammer, who was also board chair of Occidental Petroleum, notorious as a manufacturer of known carcinogens. The same industries that poison our bodies and the earth offer to brew up a cure. It's a closed circle with a benevolent public face of the sort that, during the 1960s cancer drives, sponsored cute kids like me, the one who had "lost her mother to cancer," to go door to door extorting nickels and dimes from their neighbors to "fight this deadly disease with no cure."[1]

Meanwhile, one in eight U.S. women will be diagnosed with breast cancer this year (Rennie 38). The mortality rate is higher now in most age groups than it was twenty-five years ago when my mother died of the disease. Now each year when my sister and I get our pap smears, we are warned that we are genetically disposed to "female cancers"; we are "high risk." We are encouraged to think of our breasts as enemies. The industry says that our uteruses and ovaries, too—everything contaminated by the womanly hormone, estrogen—conspire against us. Nature is the mother of a future full of horrors.

❊ ❊ ❊

African American legal scholar Patricia Williams writes about multiple categories of oppression and their intersection in explaining how she came to be interested in commerce law as an area of legal specialization.

> A few years ago, I came into the possession of what may have been the contract of sale for my great-great-grandmother. It is a very simple but lawyerly document, describing her as "one female" and revealing her age as eleven; no price is specified, merely "value exchanged . . . ." Since then I have tried to piece together what it must have been like to be my great-great-grandmother. She was purchased, according to matrilineal recounting, by a man who was extremely temperamental and quite wealthy. I try to imagine what it would have been like to have a discontented white man buy me, after a fight with his mother about prolonged bachelorhood. I wonder what it would have been like to have a thirty-five-year-old man own the secrets of my puberty, which he bought to prove himself sexually as well as to increase his livestock of slaves. I imagine trying to please, with the yearning of adolescence, a man who truly did not know I was human, whose entire belief system resolutely defined me as animal, chattel, talking cow. (17–18)

Williams's great-great-grandmother, her white owner's chattel, was forced to bear several of his children. Williams says that

> they grew up in his house, taken from her as she had been taken from her mother. They became haughty, favored, frightened house servants who were raised playing with, caring for, and envying this now-married man's legitimate children, their half brothers and sister. Her children grew up reverent of and obedient to this white man—my great-great-grandfather—and his other children, to whom they were taught they owed the debt of their survival. It was a mistake from which the Emancipation Proclamation never fully freed any of them. (18)

Various institutional forms of coercion continue to reinforce the view that women, and particularly women of non-European descent, are identical with and subject to passive matter, or the body. Anthropologist Emily Martin, who looked at the language and practices of contemporary western reproduction, asked: "If the doctor is managing the uterus as machine and the woman as laborer, is the body seen as a 'product'?" (65). Martin argues that this is the case and that women who are economically or socially disadvantaged in the society experience considerable coercion as reproductive laborers. And Susan Bordo, writing six years after Martin, looks at court-ordered obstetrical interventions and comments:

> [T]he statistics make clear that in this culture the pregnant, poor woman (especially if she is of non-European descent) comes as close as a human being can get to being regarded, medically and legally, as "mere body," her wishes, desires, dreams, religious scruples of little consequence and easily ignored in (the doctor's or judge's estimation of) the interests of fetal well-being. In 1987, the *New England Journal of Medicine* reported that of twenty-one cases in which court orders for obstetrical intervention were sought, 86 percent were obtained. Eighty-one percent of the women involved were black, Asian, or Hispanic. (76)

The deep metaphysical belief that the land does not, *cannot*, own itself is a belief from which the Wilderness Act and the Endangered Species Act have never fully freed us. And the belief that women do not own themselves is one from which the Nineteenth Amendment or even *Roe v Wade* haven't freed us, and I include many women in that "us." Why have so many pieces of progressive legislation failed? Because we haven't given up the idea of chattel. When whites were finally prohibited from killing or physically restraining human beings of African descent who walked away from the farm or the manor house, whites did not stop thinking of those human beings as objects. They were useless objects or objects that one could legally keep in a kind of economic slavery. When we passed the Wilderness Act and the Endangered Species Act, we did not start thinking of ecosystems or species as subjects. We retained the power to extract, at our discretion, certain "resources" from the guts of those "designated" wild places, the power to put them to some "use." We retained the power to eliminate species when the economic cost to us was judged by us too high, when we judged them of no utility or of negative utility. In 1973 when the U.S. Supreme Court—a court of nine males all but one of whom was white—recognized a woman's right to reproductive choice, we didn't stop viewing women as walking vaginas or wombs. This was particularly evident during the 1996 presidential election year, twenty-three years after *Roe*, as the Republican National Party was driven by the far right intent on criminalizing abortion. This was also the year that the Endangered Species Act was on the block, that salvage timber sales were running roughshod over our environmental legislation, and that takings issues were being seriously debated.

Meanwhile, the EPA reported in 1995 that it had recovered less than one-fifth of the Super Fund clean-up costs from polluters, who still owe $2.3 billion in damages to the environment. Protecting the integrity of women's bodies and of the earth are not law enforcement priorities, and this is no coincidence. We do have laws—a few laws, brave laws, precious laws—that recognize the autonomy of places and animals, the autonomy of women over their sexual and reproductive lives. But those laws are vulnerable to the Law (traditional liberal, political, and legal doctrine, precedent, process, practitioners), which has been about as friendly to these ideas of autonomy as most ranchers are to the idea of neighboring with coyotes. Within the dominant world-view such ideas are nonsense because coyotes, people of non-European descent, and women are essentially unworthy and, importantly, threaten to steal what already and rightly belongs to somebody else. That we have such laws is itself an indictment of the Law and should raise our suspicions rather than reassure us.

❊ ❊ ❊

The body as bioregion. The bioregion as me. I mean this in the literal sense and not in the more figurative sense in which many bioregionalists say this to describe their relationships with their geographical place. I don't believe that I am identical with any geographical place—with the northern Rocky Mountain ecosystem or with the Missoula Valley—or with others of those who inhabit these places: wolf, coyote, lupine, larch. Such talk, frequently heard from deep ecologists, seems arrogant. It denies others their subjectivity, the coyote her otherness, in ways dangerously similar to the way Patricia Williams's great-great-grandmother was denied hers and, more closely, in the ways my mother was denied hers. In all these cases one person risks mistaking her or his own desires for the desires of these others.

The only bioregion that we can claim strict identity with is the body. A human body is sixty electrical jolts a minute, at rest; twenty-five feet of gut, containing a virtual hothouse of microbes, each with its own diet; ninety square yards of alveoli, all performing the elegant exchange of oxygen and carbon; a mind that blips continuously up and down an eighteen-inch rope of salty brain-stuff the thickness of a man's finger. To be "home" is first to inhabit one's own body. We are each, as body, a biological ecosystem as complex, efficient, and as fragile as the Brooks Range, the Everglades, a native prairie.

Most environmentalists, including the bioregionalists, have little to say about the body; with the exception of Wendell Berry, they often say nothing about the body at all. Berry writes, "While we live our bodies are moving particles of the earth, joined inextricably both to the soil and to the bodies of other living creatures. It is hardly surprising, then, that there should be some profound resemblances between our treatment of our bodies and our treatment of the earth. . . . Contempt for the body is invariably manifested in contempt for other bodies—the bodies of slaves, laborers, women, animals, plants, the earth itself" (97, 105).

I think that we should be every bit as intelligent about our own bodies as we are about whatever geographical place we call "home." This is true of both women and men. Men and women inhabit slightly different bodily regions by virtue of, mostly, our different reproductive capacities. The bigger difference has to do with the politics of the body. Most Westernized men and women stand in a similar confused and unhealthy relationship to both their bodies and the earth, and what we do to both, with frequency, is sacrilege. I'm not just repeating here the old adage about your body being your temple. That adage reflects precisely the kind of world-view that I'm trying to debunk. It says that your body is an object that houses something *else* that is holy. I'm saying that your body is the sacred itself. Seek no further: you've found divinity in your toenails.

Some feminists express discomfort with what they read as reductionism and biological essentialism in some writing about the body, so I should qualify what I say here. I think it's possible to position oneself critically with regard to the cultural stories of the body—the raced body, the gendered body, for example—in order to invent new stories. This does not entail, necessarily, reducing women to essential bodies, to nature, as distinct from culture, or as distinct from any other of the binary categories—reason or the transcendent—that some feminists have been anxious, perhaps overly so, to claim for women.

I do think that the invention of new stories about the body will necessitate that we pay attention to our bodies, that our bodies not only will place certain limiting conditions on the stories we invent but that they also will be active narrators, if we attend to them; they will interrupt, talk back, and rupture stories when we don't attend to them. This is to leave room for plenty of variation in bodies and in the stories that we tell about them, but also to concede—as I believe some feminists who are overly enamored of language should concede—that such narratives cannot be generated out of pure imagination if they are to be of use to us. The disembodied view from the modernists' "nowhere" and from the poststructuralists' "everywhere," as Susan Bordo notes, are equally dangerous fictions (227). To be nowhere or everywhere is never to be home; to be home, inside, or even *as* the body, requires that one acknowledge the body's existence as more than a signifier, which is not to diminish the body's significance. While I want to acknowledge the profound influence of cultural inscriptions on the body, I do not want to abandon the body to a noumenal realm that is "culture." The forest that I live in is the product of some of these same inscriptions; still, every May the fairy-slipper, a rare, indigenous orchid, announces herself at the spring that also feeds my well. She is a barometer of forest health, and I should worry if she disappears. Inundated by discourses that reduce the body to a political cipher or "swaddle" her in culture, I wish to put in a word for the organism.[2]

The environmentalists' silence about the body is all too familiar. My worry is that this silence reflects that traditional and dangerous way of thinking that the body is of no consequence, that our own corporeal nature is irrelevant to whatever environmentalists are calling "Nature."

* * *

Thirty or so grizzly bears still remain in the Mission Mountains on the Flathead Indian Reservation of northwestern Montana. These bears are probably related to populations in Glacier National Park and the Bob Marshall Wilderness, to those in Idaho, and to a much larger Canadian population. Between late July and early September, the grizzlies in the Missions climb eight thousand feet onto the talus of McDonald Peak to feed on estivating populations of ladybugs. During these few weeks, the tribes close the area to human beings in an attempt to protect these bears.

On May 4, 1993, a 290-pound female grizzly bear, originally from the Missions, escaped from Wildlife Images, a "rehabilitation" center for "problem" wild animals, in Grants Pass, Oregon. This bear had been incarcerated for eating horse pellets and grain on private land and for raiding dumpsters along the South Fork of the Flathead River near Kalispell. She had been previously relocated, five times, to the eastern front of the Rockies near Great Falls, and five times she walked that several hundred miles back home to Kalispell. During this trip home she must negotiate one thousand miles and cross the Cascades. So far on this journey she's eluded men and dogs, helicopters and heat sensors, honeyed traps, guns.

One of the most remarkable things about her story is her escape from her holding pen. Conjecture has it that she climbed to the top of the pen, jimmied some welds to create an opening, grabbed an overhanging sapling by her teeth, and then heaved and squeezed all 290 pounds of herself through a hole two-and-one-half-feet long by eight-inches wide to freedom. It was impossible. The distance between the ground and that sapling was over seven feet. Her keepers don't know quite what to make of this bad bear, except that she's trouble, and that's what they've named her. For a few days the media celebrated her escape, referring to her as simply "it" or as "Trouble," befuddled by this "nonhuman subject," this conceptual outlaw. The federal authorities haven't ruled out shooting her if they find her.

I like to think of this grizzly bear giving birth to herself through that tiny opening forced by her own genius. I think of the way she trusted her weight to the muscles in her mouth, her neck, her huge humped shoulders, of how she willed her body skyward, a slurred motion opening out toward a familiar configuration of stars, distinct above the cloudy condensation of her own breath. Once she hit the ground she felt a knowledge pass through her whole body, familiar. Then she turned east and walked into the Kalmiopsis Wilderness Area and disappeared.

It is my great hope that she will reappear at McDonald Peak by August, where she will fatten herself on some of the small pleasures of home.[3]

Notes

1. I recall the American Cancer Society using this phrase in their promotional materials in the 1960s.
2. The image of the body "swaddled" in culture is Bordo's.
3. The grizzly of this last passage was shot to death by Tim O'Leary, a bear hunter, outside of Grants Pass, Oregon, on September 4, 1993, four months, minus a day, after her escape, and shortly after I wrote the first draft of this essay. When O'Leary found her she had lost forty pounds and was weakened. But she managed to fight his dogs when they attacked her and to charge O'Leary when he began shooting. This essay is dedicated to her.

Works Cited

BERRY, WENDELL. *The Unsettling of America*. New York: Sierra Club/Avon, 1973.

BORDO, SUSAN. *Unbearable Weight: Feminism, Western Culture, and the Body*. Berkeley: U of California P, 1993.

CLORFENE-CASTEN, LIANE. "Inside the Cancer Establishment." *Ms* May-June 1993: 57.

MARTIN, EMILY. *The Woman in the Body: A Cultural Analysis of Reproduction*. Boston: Beacon, 1987.

PLANT, JUDITH. "Revaluing Home: Feminism and Bioregionalism." In *Home! A Bioregional Reader*. Ed. Christopher Van Andrass, Judith Plant, and Eleanor Wright. Philadelphia: New Society, 1990. 21–23.

PLUMWOOD, VAL. *Feminism and the Mastery of Nature*. New York: Routledge, 1993.

RENNIE, SUSAN. "Breast Cancer Prevention: Diet vs. Drugs." *Ms* May-June 1993: 38–46.

RICH, ADRIENNE. *Of Woman Born*. New York: Norton, 1979.

WILLIAMS, PATRICIA. *The Alchemey of Race and Rights*. Cambridge: Harvard UP, 1991.

PART III
READINGS OF NINETEENTH-CENTURY ENVIRONMENTAL LITERATURE

The Ornithological Autobiography of John James Audubon
CHRIS BEYERS

From 1827 to 1840 John James Audubon published *The Birds of America*, an audacious series of volumes that attempted to depict every bird native to the United States *life-size*. Less well known is Audubon's five-volume companion work, *Ornithological Biography* (1831–40), prose descriptions of all the birds illustrated in *The Birds of America*. Though his popular reputation rests upon his accomplishments as a visual artist, the naturalist wrote a great deal of impressive prose besides *Ornithological Biography*. Audubon's literary output includes a substantial collection of descriptions of animals, anecdotes of frontier life, journals, memoirs, and miscellaneous articles. Audubon's writing is especially valuable because it offers something generally absent in his illustrations: a portrait of the naturalist himself.

Unfortunately, many of the salient details of this portrait are difficult to find. Since Audubon's first language was not English, he relied on others to correct his writing for him. Audubon's editors have often taken this as sanction to edit content as well, and twentieth-century editors have often attempted to mold Audubon into the naturalist they would like him to have been. For example, in one of her many editions of Audubon's prose, Alice Ford assures readers that she expunged "cruel" passages that might offend "champions of the cause of conservation and wildlife" (vi). One of the best handy editions of Audubon's prose, Scott Russell Sanders's *Audubon Reader*, avoids such bowdlerization. However, as a selection (its subtitle is *The Best Writing of John James Audubon*), Sanders's edition necessarily omits important material.

Sanders's principles of selection are suggested by the introduction to the volume, where he quotes what he calls an "archetypal moment" in which Audubon is "delicately balanced among four rival identities": hunter, artist, scientist, and nature lover. "What makes Audubon extraordinary, and what charges these writings with inner drama," Sanders continues, "is the *fusing* of these roles, the fierce interplay of identities" (7). Unfortunately, most editions make it difficult to get a good view of all of Audubon's "identities," and readers have too often used incomplete evidence when appraising the Audubon who appears in the prose. This article uses some lesser-known passages—ones that rarely, if ever, appear in anthologies of Audubon's prose—in order to give a more accurate and complete description of the self-portrait Audubon provides in his writings. In particular, I will focus on those points at which Audubon was most *un*balanced, times when one of his roles interfered with another. I will also examine what Audubon thought about the contradictions he discovered in his own approach to the natural world.

The "four rival identities" Audubon attempted to balance are evident in two of his accounts of why he began drawing birds in the first place. Because these accounts constitute an exemplary tale showing how Audubon *could* fuse his various roles, they are a good place to begin an examination of the way conflicts often prevented those roles from being harmonized.

From his early youth, Audubon says, he felt an "intimacy" with nature, one "bordering on phrenzy." Although many praised his first drawings, his father was not impressed, explaining that living things were not easy to imitate. Audubon *père* did, however, encourage his son to study birds so the young man might "raise" his "mind toward their great Creator" (Audubon 1:vi). Audubon describes his first wildlife drawings as "stiff, unmeaning profiles" (qtd. in Ford 11).

After moving to America, Audubon tried again, this time by shooting birds and then hanging them by a string to make them appear lifelike. "In this manner," he remarks, "I made some pretty fair signs for poulterers." Audubon next returned to the outdoors, simply "outlining" birds as he observed them. Since the birds would not stay still for minute observation, he found he could "finish" none of his sketches. Consequently, he then killed a number of birds and laid them on a table before him. He found, not surprisingly, that "they were all *dead*" (qtd. in Ford 11), and "The moment a bird was dead . . . it could no longer be said to be fresh from the hands of its Maker." Audubon longed to capture "Nature, alive and moving" (qtd. in Ford 11).

Finally, Audubon hit upon the idea of mounting the birds on a piece of wood and then using wires to pose them naturally. The first illustration of a bird so mounted, he remarks, was "what I shall call my first drawing actually from Nature, for even the model's eye was still as if full of life when I pressed the lids aside with my finger" (qtd. in Ford 12). Only one problem remained: he wasn't sure what, precisely, was the bird's natural attitude; so off he went, back to nature, to study avian habits. "The better I understood my subjects," he explains, "the better I became able to represent them" (qtd. in Ford 13).

This narrative of how Audubon learned to represent nature shows him at his most hopeful. Aesthetic pleasure is consonant with both spiritual fulfillment and scientific inquiry. In order to understand the world and appreciate it fully, Audubon became a frontiersman to explore it, a scientist to observe it, a nature lover to feel it, and an artist to represent it. The narrative illustrates the way that Audubon could integrate his various impulses; it is a naturalist's version of what Matthew Arnold would call Wordsworthian joy.

Audubon, though, could not always maintain this balance. At times, for example, the aesthetic feelings of the nature lover interfered with the naturalist's scientific aims. This is evident in his descriptions of bird calls. Normally, Audubon's representations of the calls of birds are precise and inventive: the Booby Gannet's song is "harsh and guttural, somewhat like that of a strangled pig, and resembling the syllables *hork, hork*" (3:64), while the Ivory-Billed Woodpecker's notes "resemble the false high note of a clarinet [sic]. They are usually repeated three times in succession, and may be represented by the monosyllable *pait, pait, pait*" (1:343). But when he finds a bird's song truly beautiful, he abruptly loses his powers of description. "Would that I could describe the sweet song" of the Fox-Coloured Sparrow, he exclaims. The song offers a "delightful serenade." He describes it as "enchanting" and "powerful." "But, reader," he concludes, "I can furnish no description of the melody" (2:59). Many times in *Ornithological Biography* Audubon finds himself similarly at a loss for words when encountering natural music. Like a

Renaissance sonneteer speaking about his beloved, Audubon's rhapsodies focus on the state of his own mind and not on the object that inspired him, thus providing superlatives but no real description.

Similarly, Audubon's empathy with natural creatures causes him to assert that birds have feelings analogous to, if not the same as, those of humans. He often tells the reader that a bird is "courageous" or "cowardly," "vicious" or "gentle," "solicitous" or "tyrannical." Birds mate, he seems to think, because they fall in love and sing because they are happy. Such emotional identification with nature blinds him to more scientific analyses—for example, the possibility that birds sing to establish territory.

Audubon's inability to fuse his roles has more severe consequences than simply preventing him from writing precise descriptions. A much more serious conflict is implicit in the tale of how he learned to depict birds. Although inspired by a great love of birds, he finds he must kill them to represent them accurately. It's an open secret among Audubon scholars that the great ornithologist killed thousands of birds, including literally hundreds of a single species, such as the Chimney Swallow, the Lapland Lark-Bunting, and the now-extinct Passenger Pigeon. Indeed, a typical entry in *Ornithological Biography* describes not only a bird's plumage, song, and mating habits, but also offers an account of how it behaves after it has been shot, and what it tastes like once cooked.

Most readers have assumed that Audubon rather blithely killed wildlife and that he felt his depredations fully justified. For example, in the introduction to Ben Forkner's recent collection of Audubon's prose, Forkner remarks that modern readers may be put off by the fact that the naturalist killed large numbers of birds "without the slightest sign of guilt" (xxi). Barbara J. Cicardo enlarges upon that idea, commenting that Audubon's "acknowledgment of man's right to behave in such a brutish fashion is part of [a] larger vision of his place in the natural history of things, an almost fatalistic vision which both Robert Penn Warren and Eudora Welty . . . use in their fictional creations of the character Audubon" (88). Cicardo's observation not only accurately describes the way Warren's *Audubon: A Vision* and Welty's "A Still Moment" portray the naturalist, but also the way most critics seem to understand him. Although Annette Kolodny acknowledges Audubon's anxieties about his actions, she argues that when Audubon kills the animals he admired he demonstrated that he was "unable to resolve" the central paradoxes of his existence (88). Audubon settled, she argues, on a world-view equating knowledge with violence—one that vindicated his masculine desire to dominate. Not so, argues Elisa New; Audubon's art (both visual and literary) "unites disparities" (404) and allows us to look at "creation's strangely electric, because ineluctably lethal, vitality" (403).

All these views assume that Audubon's *Ornithological Biography* either granted or sought to grant the naturalist the "right" to kill. However, some passages suggest otherwise. Though at times Audubon exhibits no remorse over killing birds (his entries on many kinds of ducks amount to little more than hunting directions), at other times doubts dog him. Consider his description of the Clapper Rail, in which Audubon tells his reader that hunting the animal is "exceedingly pleasant" and then describes a typical outing. Early in the morning his party of hunters goes to the marshes, begins shooting, and is soon joined by many other hunters. When the shooting ends, Audubon looks around:

> It is a sorrowful sight, after all: see that poor thing gasping hard in the agonies of death, its legs quivering with convulsive twitches, its bright eyes fading into glazed obscurity. In a few hours, hundreds have ceased to breathe the breath of life; hundreds that erst revelled in the joys of careless existence, but which can never behold their beloved marshes again. The cruel sportsman, covered with mud and mire, drenched to the skin by the splashing of the paddles, his face and hands smeared with powder, stands amid the wreck he had made, exultingly surveys his slaughtered heaps, and with joyous feeling returns home with a cargo of game more than enough for a family thrice as numerous as his own. (3:37)

Clearly distressed, Audubon asks the hunter how he justifies the "cruelty of destroying so many of these birds." The hunter replies, "It gives variety to life; it is good exercise, and in all cases affords a capital dinner, besides the pleasure I feel when sending a mess of Marsh-Hens to a friend such as you" (3:38). The entry ends here without indicating if the reader is expected to accept this justification. If Audubon were attempting to resolve inconsistencies, we might expect him to explain which point of view is to be believed and why.

Audubon often implies very strongly that he does not approve of killing animals, especially not in a manner that is cruel or unnecessary. For example, his often-anthologized description of the slaughter of Passenger Pigeons portrays the event in infernal terms. More telling, perhaps, is "The Red-Tailed Hawk," in which Audubon tells of a hawk that, "impelled by continued hunger" seizes a hen (1:266). The chicken's owner hears the "dying screams" of the chicken, and "swears vengeance." He follows the bird to its nest and promptly begins to chop down the tree. Though the hawk "sails sorrowfully over and around," and "would fain beg towards mercy for her young," the farmer eventually shoots the bird and makes considerable progress in felling the tree (1:266).

> The huge oak begins to tremble. Were it permitted to speak, it might ask why it should suffer for the deeds of another; but it is now seen slowly to incline, and soon after with an awful rustling produced by all its broad arms, its branches, twigs and leaves, passing like lightning through the air, the noble tree falls to the earth, and almost causes it to shake. The work of revenge is now accomplished: the farmer seizes the younglings, and carries them home, to be tormented by his children, until death terminates their brief career. (1:267)

Characteristically, Audubon does not ignore the pain of any of the actors in his drama. The separate agonies of chicken, hawk, and tree are all depicted clearly. However, while the hawk is impelled by hunger, the farmer's motivation is "revenge," and torturing the bird's young is not necessary for his survival. The diction suggests that the farmer's action imperils more than his own soul. Audubon's prose invokes the traditional symbolic value of the oak as a tree that stands for continuity and strength, and when this "noble tree" falls, it "almost causes" the earth "to shake" (1:267). Symbolically, the farmer's actions threaten the world's moral stability.

Why killing a bird should so shake the world becomes clear in passages where Audubon recognizes his *own* culpability for depredation. In "The Mississippi Kite," for instance, he tells of happening upon the nest of the Kite, a species he had not seen before. Desiring closer study, he shot at the bird but missed; the bird caught sight of him, but nevertheless proceeded to feed her young "with great kindness." Audubon again shot and missed, and this time the kite "gently lifted her young, and sailing with it to another tree, about thirty yards distant, deposited it there" (2:109). The bird's valor thrilled Audubon: "My feelings at that moment I cannot express. I wished I had not discovered the poor bird; for who could have witnessed, without emotion, so striking an example of that affection which none but a mother can feel; so daring an act, performed in the midst of smoke, in the presence of a dreaded and dangerous enemy." What did Audubon do next? He walked over and shot the bird and her young. "So keen," he remarks, "is the desire of possession!" (2:109).

Audubon's entry on the Arctic Tern also expresses his sense of guilt about killing what he loves. He found that the tern's "easy and graceful motions" provoked in him "a desire to possess it." Since he had made notes on their habits, he needed to shoot some of the birds to "finish his picture from life" (here again, when Audubon says "life" he means "death"). He continues, "Alas, poor things! how well do I remember the pain it gave me, to be thus obliged to pass and execute sentence upon them. At that very moment I thought of those long-past times, when individuals of my own species were similarly treated; but I excused myself with the plea of necessity, as I recharged by double gun" (3:367). Audubon might have reflected that "necessity" was and often is the excuse his own species uses to pardon its killing of itself. As in "The Clapper Rail," it is unclear whether the reader is expected to accept Audubon's rationalization, and Audubon finds himself again driven to become executioner by the desire for "possession"—the same evil to which Adam and Eve fell prey.

What rendered the killing of birds especially troubling to Audubon was his belief that humans reside in a significant universe, and he often asserts his pre-Darwinian notions of teleology. For example, in "The Cow-Pen-Bird," he tells his reader that "the manifestations of consummate skill everywhere displayed" in nature lead us to "infer that the intellect that planned the grand scheme, is infinite in power" (1:493), thus echoing the similar sentiments of his father in the narrative of how he learned to draw birds. In fact, Audubon's anthropomorphic tendencies coupled with his moral sense sometimes led him to argue that nature provides plainer lessons for humanity. For example, he uses his account of the Blue Jay, which has "resplendent" plumage but acts with "selfishness, duplicity, and malice," to show how appearances can be deceiving (2:11). He uses a picture of two Ferruginous Mocking Birds defending their nest to "remind" readers that "innocence, though beset with difficulties, may, with the aid of friendship, extricate herself with honour" (2:102). Clearly, Audubon sometimes read nature as an emblem-book.

The natural world, then, offers humans a glimpse of both the Creator's plan and their own character. And as the "The Mississippi

Kite" and "The Arctic Tern" demonstrate, nature is sometimes the stage for a morality play in which fallen beings interact with an American Eden. It is worth noting that Audubon's belief that humans are fallen from grace needn't be read as his way of condoning brutality; the recognition that people are prone to sin is traditionally used as an explanation for human misdeeds, not as an excuse.

Audubon understood nature as both blueprint of the world's design and as an opportunity for personal redemption. In his entry on the Wood Thrush, for example, he describes himself caught in a violent, nighttime storm which shatters "the stateliest and noblest tree in my immediate neighborhood." As in "The Red-Tailed Hawk," the fallen tree suggests that the world's stability has been shaken. After the storm ends, he discovers that his notes and drawings are destroyed and that he is "shivering in a cold fit like that of a severe ague." He begins to despair that he has wasted his time and that he will never see his family again. Suddenly he hears the song of the Wood Thrush, coming "as if to console me amidst my privations, to cheer my depressed mind, and make me feel, as I did, that never ought man to despair, whatever be his situation, as he can never be certain that aid and deliverance are not at hand" (1:372–73). More extravagantly, "The Zenaida Dove" relates the dubious story of how the cooing of the dove persuaded a pirate to leave his life of plunder and return to his family. In the same entry, however, Audubon dismisses fishermen's accounts of the migratory patterns of the dove because he had not seen the migrations for himself. The account suggests a distinctive paradigm: in matters of observation, Audubon was a strict empiricist; in matters of interpretation, he was willing to believe.

Although there is considerable evidence to support the contention that Audubon could sometimes balance his various roles, and that he was capable of even cold-hearted killing, there is also substantial evidence that he was often unable to maintain equipoise, and that this troubled him. Audubon often appears in his writings as a naturalist keenly conscious of the contradictions in his own behavior. Thus, Sanders's notion that Audubon "fused" his various roles, or that he "united disparities," explains only part of the story. When Audubon's empirical eye helps him see the harmonious design his nature-loving side sought, his roles do fuse. But when he states two contradictory notions one after the other without comment, or when he articulates an attitude but acts in contradiction to it, it is difficult to understand how this represents a unification of his roles. It seems that Audubon's identities were often as divided as they were fused.

My intention here is not to quibble about what we mean when we say identities are fused, but rather to point out a prevalent misreading of Audubon's writing. The naturalist has frequently been read as if his prose were didactic and exemplary. This is why Ford, for example, sees fit to excise passages that disgust her—she does not want anyone to emulate what she considers reprehensible actions and attitudes. Though Audubon is seldom shy about dispensing advice, he almost never wrote an essay in the manner that I am writing one now, in a way that constructs arguments and draws conclusions. Instead, his writings have an off-handed, occasional air. Audubon writes journals, letters, "episodes"—written, he explains, to "relieve the tedium" (1:29)—anecdotes, memoirs, entries, and the like. His genius was as a descriptive writer, and that is how his prose ought to be read.

What is striking about Audubon's frank acknowledgments of his guilt over his own depredations is his dramatic presentation. He brings the reader to the instant when the naturalist surveys the bloodshed, and then, after a phrase like "at that moment," he simply describes the thoughts that came into his mind and his subsequent actions. Audubon isn't really *telling* us anything, but he is *showing* us something significant. His descriptions do not constitute a philosophy constructed from precise arguments; instead, they simply relate events as they transpired.

Still, I suspect few twentieth-century readers of Audubon do not feel at times that Audubon is something of a hypocrite. After all, if he loved these animals so much, why did he continue killing them? It cannot be argued that Audubon held a philosophy that condoned his own behavior, because he frankly admits he does not. Indeed, he frequently articulates a system of ethics that would damn many of his own actions. But in this contradiction I find something very familiar. When Audubon champions the cause of nature but proceeds to kill wildlife, I wonder if his contradictions are considerably worse than our own. I try to walk places when I can, but if it is raining, or I have a long distance to go, or I'm tired or in a hurry, I take my car—despite the fact that I know that it burns fossil fuels, destroys the atmosphere, and contributes to the congestion on the roads, which leads to habitat destruction when new roads are built to relieve the congestion. Even the production of the volume in which this essay appears consumed precious resources, and it is not entirely clear to me that those resources would have been better left untouched. Like Audubon, I proceed from a vague and uneasy notion that what I am doing will, in the long run, help more than hurt the cause I espouse. However exemplary or contemptible we may find him, Audubon is a complex character whose competing impulses and instincts dramatize conflicts with which we continue to struggle today.

Works Cited

AUDUBON, JOHN JAMES. *Ornithological Biography, Or an Account of the Habits of the Birds of the United States of America; Accompanied by Descriptions of the Objects Represented in the Work Entitled Birds of America, and Interspersed with Delineations of American Scenery and Manners.* 5 vols. Edinburgh: Adam & Charles Black, 1831–39.

CICARDO, BARBARA J. "From Palette to Pen: Audubon as Writer." *Audubon: A Retrospective.* Ed. James H. Dormon. Lafayette: U of Southwest Louisiana P, 1990. 74–91.

FORD, ALICE. *Audubon, by Himself.* Garden City: Natural History P, 1969.

FORKNER, BEN, ED. *John James Audubon: Selected Journals and Other Writing.* New York: Penguin, 1996.

KOLODNY, ANNETTE. *The Lay of the Land: Metaphor as Experience and History in American Life and Letters.* Chapel Hill: U of North Carolina P, 1975.

NEW, ELISA. "Beyond the Romance Theory of American Vision: Beauty and the Qualified Will in Edwards, Jefferson, and Audubon." *American Literary History* 7.3 (1995): 381–414.

SANDERS, SCOTT RUSSELL, ED. *Audubon Reader: The Best Writing of John James Audubon.* Bloomington: Indiana UP, 1986.

"*A beautiful and thrilling specimen*"
George Catlin, the Death of Wilderness, and the Birth of the National Subject

DAVID MAZEL

> *Nature as the self's (flattering) mirror, but not ever, no never, Nature-in-itself.*
>
> JOYCE CAROL OATES (464)

Though George Catlin went West primarily to paint portraits of Native Americans, he also took a keen interest in the landscapes and wildlife he encountered—so much so that his now-classic *Letters and Notes on the North American Indian* (1841) reads at times like an early environmentalist manifesto. He laments at some length, for example, the "wasting" of that "noble animal," the buffalo, taking care to link the disappearance of the bison to the demise of the Plains Indians, whose existence depended so much upon it. In direct response to this "melancholy contemplation," this felt need for government intervention on behalf of endangered wildlife *and* endangered people, Catlin makes the first recorded appeal for an American national park:

> [W]hat a splendid contemplation . . . when one . . . imagines them as they *might* in future be seen, (by some great protecting policy of government) preserved in their pristine beauty and wildness, in a *magnificent park*, where the world could see for ages to come, the native Indian in his classic attire, galloping his

> wild horse, with sinewy bow, and shield and lance, amid the fleeting herds of elks and buffaloes. What a beautiful and thrilling specimen for America to preserve and hold up to the view of her refined citizens and the world, in future ages! A *nation's Park*, containing man and beast, in all the wild and freshness of their nature's beauty! (261–62)

In its invocation of the beautiful and the sublime, in its sense of awe and urgency and its suggestion of concrete political action, Catlin's writing strikes me generally as familiar and right. It is the product of a recognizably environmentalist sensibility, and I think Catlin is rightly praised as a prophet of the wilderness preservation movement. But his union of wild landscapes and "wild" people strikes me as decidedly problematic, for the notion of rendering an entire people into a living tableau is clearly the product of an *imperialist* sensibility, while the idea of displaying such a "specimen" to an admiring world just as clearly reflects a *nationalist* sensibility.

In this essay I wish to explore the conceptual and functional links between Catlin's environmentalism, his imperialism and nationalism, and the creation of an American identity. I am aware, of course, that Catlin's attitudes can be seen as humane in light of the subsequent history of both the environment and the Native Americans. Yosemite Valley, to give an example, the nation's first great wildland preserve, would soon be "discovered" and bloodied in the very process of hounding native peoples out of it, and this particular instance typifies the more general history in which America's remaining wilderness would be organized into a system of *preservation* and its surviving natives into the decidedly different system of the *reservation*. My purpose, however, is not to gauge Catlin's morality, but to ponder his language. What is one to make of his discursive linkage of people and places that were in practice to be so violently separated? Is this linkage fundamental, or merely coincidental, or again "merely" purposeful?

One could suggest that in Catlin's case it is the latter, that he disguises an otherwise purely environmentalist sensibility in an imperialist rhetoric cleverly designed to appeal to the prejudices of his contemporary readership. Consider, for example, his observation that wherever the buffalo are found in abundance, the Plains Indians are self-sufficient and peacefully preoccupied, while where the buffalo are absent, the native is both an expense to the government and a danger to the citizenry:

> [W]hen the buffaloes shall have disappeared in his country, which shall be within *eight* or *ten* years, I would ask, who is to supply the Indian with the necessaries of life then? and . . . when the skin shall have been stripped from the back of the last animal, who is to resist the ravages of 300,000 starving savages; and in their trains, 1,500,000 wolves, whom direst necessity will have driven from their desolate and gameless plains, to seek for the means of subsistence along our exposed frontier? (263)

Again proposing what is at once an environmentalism *and* an Indian policy, Catlin shrewdly plays to the ingrained fear of Indian resistance along the "exposed frontier," drawing intentionally, I think, upon the long-established conceptual association of wildlands with "wild" people.

But even if that linkage seems consciously exploited in this example, it may yet inhere in the language itself. And it is this possibility—that Catlin in effect had little choice, that the discourse of wilderness which he had inherited *precluded* any discussion of wildlands without the simultaneous invocation of ideas about wild people—that most fundamentally implicates his environmentalism in the national and imperial politics of his time.

Though they were not identical, the discourses of the "savage" and of the "wilderness" were functionally aligned within an overarching discourse of "civilization." In *The Savages of America*, Roy Harvey Pearce has examined "what it meant for civilized men to believe that in the savage . . . there was manifest all that they had long grown away from" (ix). In America, he concluded, such "civilized men . . . could survive only if they believed in themselves," and up until the mid-nineteenth century,

> that belief was most often defined negatively—in terms of the savage Indians who, as stubborn obstacles to progress, forced Americans to consider and reconsider what it was to be civilized and what it took to build a civilization. Studying the savage, trying to civilize him, destroying him, in the end they had only studied themselves, strengthened their own civilization, and given those who were coming after them an enlarged certitude . . . in the progress of American civilization over all obstacles. (ix)

The Indian, in other words, was one term in a dialectic through which Euro-Americans might link an expanding state power with a nascent national identity. In *Wilderness and the American Mind*, Roderick Nash makes a similar claim—not about the Indian, however, but about the wilderness, which "was the basic ingredient of American civilization. From the raw materials of the physical wilderness Americans built a civilization; with the *idea* or *symbol* of wilderness they sought to give that civilization *identity* and *meaning*" (xv, emphases added). Pearce and Nash distinguish between, on the one hand, the literal dispossession of real Indians and the physical exploitation of their real lands, and, on the other hand, the conceptual deployment of an idealized Other, an actor cast in a psychohistorical drama of American identity. To the extent that the tangible, historical realities of native peoples and landscapes failed to accord with the script, they were more or less irrelevant in this drama. Such realities were in fact *obstacles* to be overcome not with plows and guns but with words, through the discursive construction of an uncivilized and ultimately powerless Other against which to define a civilized and seemingly omnipotent Self.

Legitimizing America
Wilderness as National Resource

> *The wilderness is at once secured and obliterated by the official gestures that establish its boundaries.*
>
> STEPHEN GREENBLATT (9)

This entanglement of people, place, identity, and power is evident in the very etymology of "wilderness." A term "heavily freighted with meaning of a personal, symbolic, and changing kind," as Nash notes, wilderness is not easy to define—even on the most practical level, where "land managers and politicians have struggled without marked success to formulate a workable definition" (5). It cannot be defined simply by enumerating its components, for "the number of attributes of wild country" seems to be "almost as great as the number of observers" (1). This subjective aspect of the term, which precludes any "universally acceptable definition," stems from the fact that "while the word is a noun it acts like an adjective. There is no specific material object that is wilderness. The term designates a quality (as the '-ness' suggests) that produces a certain mood or feeling in an individual and, as a consequence, may be assigned by that person to a specific place" (1). The term "wilderness" is like "environment" in suggesting senses of activity and passivity that are obscured in everyday usage. It is difficult to delineate any "specific material object" that is "the environment," and this is so because the environment (as the "-ment" suggests) is a nominalized verb. Though today used exclusively to designate a thing, the term is rooted in an action, in the verb "to environ" (or "surround"), and to speak of having or being in an environment implies that the environment has actively *environed* the speaker. But of course the environment's activity here is purely grammatical; the real action was performed by the speaker, who at some point *entered* the environment. In history if not in grammar, it makes little sense to think of being passively "environed" through some mysterious agency of the landscape, for historical agency inheres in people, not things.

Thus, however much the noun "environment" may imply action on the part of the landscape, the word points ultimately to a speaker whose environment consisted of historically specific acts of entry. It would in this sense be disingenuous to claim that Euro-Americans "were environed" by the North American environment; rather, that environment became theirs because they quite actively penetrated it and forcefully claimed (and continue to claim) it as their own. (For a detailed discussion of the etymology and politics of the term "environment," see my article "Environmentalism as Domestic Orientalism" [Mazel 138–42].) In much the same way, the noun "wilderness" points to both action and speaker. It is said to *produce* a certain mood or feeling, but of course this action of "bewildering" cannot be said to originate in the landscape, which again lacks genuine agency; instead it reflects a historically contingent mood of the speaker that has been projected back onto nature. Just as "environment" misnames an imperial penetration as a landscape, so "wilderness" misnames a cultural anxiety as a geography.

And just what is this anxiety? To be "bewildered" is to find oneself without the means to choose between a confusing array of "conflicting situations, objects, or statements," and the danger of bewilderment—the danger that echoes through the early etymology of wilderness—is the possibility that, lacking proper guidance, one might stray (from Old English, *wilder*) from the proper path. "[C]onceived as a region where a person was likely to get into a disordered, confused, or 'wild' condition," writes Nash, the key image is "that of a man in an alien environment where the civilization that *normally orders and controls his life* is absent" (2, emphasis added). Thus does this early notion relate wilderness to the contemporary Foucauldian conception of *power*, for its defining absence is precisely the absence of that pervasive complex of signs and institutions which order, control, and establish norms of thought and behavior—that is, of power as it produces the citizen of the modern state.

But the wilderness of Catlin's proposed national park is hardly "bewildering" in this sense. What Catlin wished to "hold up to the view" of America's "refined citizens" was perceived not as disorienting but as *edifying*, as reinforcing and consolidating a distinctly national identity. This is the sense, I think, of Nash's claim that "by the middle decades of the nineteenth century wilderness was recognized as a cultural and moral resource": invested with a certain set of symbolic meanings, the wilderness could serve to allay the very anxieties it had once provoked (67). Engendering what Pearce termed an "enlarged certitude . . . in the progress of American civilization over all obstacles," the wilderness could serve an ideological function, legitimizing the American state by making its expansion seem inevitable and natural, rather than merely provisional (ix). Catlin's particular genius was to see that wilderness could accomplish this cultural work not through its conquest, but through its preservation and perpetual display, that the wilderness could be brought into the service of power by the very policy of appearing not to change it at all.

This is the sense in which I speak in my title of the "death" of a wilderness, whose preservation is paradoxically its evisceration. Such a wilderness will not be tamed physically but discursively. It will become the cornerstone of an environmentalism that is not only a *de jure* land policy but also, as a discourse of subjectivity, a *de facto* people policy. Through the particular entanglement of people, place, identity, and power evinced in Catlin's writing, wilderness will function as part of the larger disciplinary apparatus of the young American republic.

Constructing Subjects
Ecology + Identity = Ecologicality

> *The wild is not to be made subject or object . . . it must be admitted from within, as a quality intrinsic to who we are.*
>
> GARY SNYDER (181)

Catlin's call for a "wild" national park signals one beginning of what I would call an American "ecologicality." By this term I refer to the processes by which environmentalism, in all its various manifestations—the love of nature, wilderness adventure, ecological awareness, and so on—has helped discipline the modern national subject. To trace out the history of those processes would require a book; what I would like to do in the remainder of this essay is to briefly examine two facets of contemporary ecologicality. First, I would like to consider how, without being explicitly named as such, ecologicality has already been acknowledged and explored. Second, I would like to take up the work of Bill McKibben, a prominent and influential ecocritic who, in my view, profoundly *mis*recognizes ecologicality. This in turn will entail an examination of how wilderness has functioned ideologically to foster that misrecognition. Catlin remains central to the discussion, although the historical context is no longer the incipient nationalism of the early nineteenth century but the dawning globalism of our own day.

Thus far, what I am calling ecologicality has perhaps been most prominently and vigorously analyzed by such critics as Andrew Ross and Joni Seager—though they have not, to my knowledge, made any effort to generalize their observations on environmental discourse and subjectivity (nor are they generally considered ecocritics).[1] More interesting for my purposes here are critics who have begun to theorize ecologicality as such. I would argue that Mitchell Thomashow is doing just that in his *Ecological Identity*, where he forthrightly claims that today's environmentally aware subject uses "environmental values to construct a personal identity" (xi). This distinctively *ecological* identity is an "orientation and sensibility" that

> involves a reconstruction of personal identity, so that people begin to see how their actions, values, and ideals are framed according to their perceptions of nature. . . . Ecological identity refers to how people perceive themselves in reference to nature, as living and breathing beings connected to the rhythms of the earth, the bio-geochemical cycles, the grand and complex diversity of ecological systems. (xiii)

An unabashedly political construct—equally responsive, I would argue, to the dictates of the global ecology and the global economy—ecological identity "is intrinsic to contemporary environmentalism" (xiv). It is also intrinsic to the contemporary subject, being, as Thomashow puts it, no less than "a lens through which the experiences of everyday life take on new meaning" (17)—something that could also be said of such other discursively constructed "-alities" as sexuality and nationality. It deftly reconfigures the old conceit of America as Nature's Nation, rearticulating it through the individual consciousness, grounding a mode of "ecological identity work" whose aim is to "forge a concept of *ecological* citizenship" (xvii, emphasis added) out of the old, merely *national* citizenship. The subject is being refashioned from the inside out, in a process that in the long run may preserve the global ecology but in the short run, as Andrew Ross insists, will serve the global economy.

Another prominent ecocritic who has begun to theorize the links between environmental discourse and contemporary subjectivity is Scott Slovic. In *Seeking Awareness in American Nature Writing*, Slovic contends that our best nature writers "are not merely, or even primarily, analysts of nature or appreciators of nature—rather, they are students of the human mind," preoccupied "with the psychological phenomenon of 'awareness'" (3). Such students of the mind are fascinated by the natural environment—the wilderness in particular—because they can enlist it precisely as Catlin implied, as the constitutive Other of the Self:

> Both nature and writing . . . demand and contribute to an author's awareness of self and non-self. By confronting face-to-face the separate realm of nature, by becoming aware of its otherness, the writer implicitly becomes more deeply aware of his or her own dimensions. . . . It is only by testing the boundaries of self against an outside medium (such as nature) that many nature writers manage to realize who they are. [2]

Slovic argues that nature writing is "a 'literature of hope' in its assumption that the elevation of consciousness may lead to wholesome political change" (18). Certainly environmentalism will get nowhere in the absence of an informed consciousness—but insight may also be a form of blindness, and there is always the danger that such an "elevation" of consciousness will wind up producing the sort of false awareness we term ideology. How does contemporary environmental discourse function as ideology? Thomashow provides an example when he writes of how, in a sort of epiphany, it "occurred" to him "that ecological identity work can occur *even in a supermarket*" (177, emphasis added). I want to focus for a moment on the sense of surprise registered by that word, *even*. I suggest that what Thomashow is taking for granted here is the preeminence of wilderness as the site of ecological identity work: "common sense" tells us that such work takes place outdoors rather than indoors, in the woods rather than in town. But why *shouldn't* ecological identity work proceed just as insightfully in a supermarket? After all, the supermarket is a complex and instructive nexus of energy flows, as pedagogically sound a window into ecological relationships as a pristine forest or wetland. And doesn't genuine environmental reform depend as much or even more upon changing our social behavior in the market as on understanding the natural state of the wilderness? Why is it that the supermarket's considerable ecological significance is so unobvious it must "occur" to the environmentalist in what seems like a flash of insight?

What the surprised tone registers here is precisely the way the concept of a pure and untrammeled "nature," seemingly the foundation of ecological awareness and identity, can itself be mystifying. Thomashow notes correctly that "[e]cological identity work requires the ability to overcome both internal and external distractions" (179), but clearly we must not dismiss as "distractions" the very social complexities which a genuinely ecological consciousness ought to keep in view. Otherwise, ecologicality operates unwittingly to provide an escape from such complexity (much as the wilderness has traditionally served as an escape from the city)—thereby rendering ecologicality profoundly anti-ecological.

Constructing Wilderness
Reality - History = Nature

> One of the most striking proofs of the cultural invention of wilderness is its thoroughgoing erasure of the history from which it sprang.
>
> WILLIAM CRONON (79)

Thomashow claims that "[a]s we cultivate ecological identity" we "are no longer satisfied to live in forgetfulness and denial" (205). Yet the American wilderness has itself been constructed as a site of forgetfulness and denial, a site from which history has been erased in order to construct a powerful but illusory vision of nature. To see how this happens—and to bring the discussion back to that pivotal figure in the history of American ecologicality, George Catlin—I wish to turn now to Bill McKibben's widely praised *The End of Nature*. Published as a book in 1989 after being serialized in *The New Yorker*, *The End of Nature* is one of the most thoughtful, accessible, and influential of the recent exegeses on the problematics of "nature." For all its insight, however, the book repeatedly mystifies the role of wilderness in ecologicality.

McKibben argues, among other things, that we live in a "postnatural" age, that the ubiquity of environmental change wrought by global industrialism now calls into question the very *idea* of nature. Untouched wilderness, as nature's conceptual ground, no longer exists—not even in as remote a place as Antarctica, whose once-pristine snowpack has been sullied with traces of wind-borne chemicals and whose very climate has been altered by global warming. Thus we can no longer construct wild nature as we once did, which is to say, by means of "[o]ur ability to shut the destroyed areas from our minds, to see beauty around man's degradation" (57): "If the ground is dusty and trodden, we look at the sky; if the sky is smoggy, we travel someplace where it's clear; if we can't travel to someplace where it's clear, we *imagine* ourselves in Alaska or Australia or some place where it is, and that works nearly as well" (58). McKibben rightly highlights how the escape to the natural has functioned as an aid to forgetting the omnipresence of the cultural, the ubiquity of our own footprints. Throughout the history of the United States, he argues further, such an escape has always been available—that is, until recently.

McKibben demonstrates that such an escape is decidedly ideological, an element in a more general process not so much of experiencing natural realities as of repressing social realities. Consider his account of the Adirondack lake in which he and his wife like to swim. "A few summer homes cluster at one end" of this lake, "but mostly it is surrounded by wild state land" (49).

> During the week we swim across and back, a trip of maybe forty minutes—plenty of time to forget everything but the feel of the water around your body and the rippling, muscular joy of a hard kick and the pull of your arms.
>
> But on the weekends, more and more often, someone will bring a boat out for waterskiing, and make pass after pass up and down the lake. And then the whole experience changes, changes entirely.

This lamentable change, McKibben makes clear, is not so much physical as mental. As important as any concrete damage to the environment is the motorboat's intrusion into an otherwise carefully guarded *psychic* territory:

> Instead of being able to forget everything but yourself, and even yourself except for the muscles and the skin, you must be alert, looking up every dozen strokes to see where the boat is, thinking about what you will do if it comes near. It is not so much the danger.... It's not even so much the blue smoke that hangs low over the water. It's that the motorboat gets in your mind. You're forced to think, not feel—to think of human society and of people. (49)

Despite the presence of the summer homes, McKibben can experience this liminal setting as wild so long as he can perform this crucial *forgetting* "of human society and of people." But it ceases to be wild when its cultural inflection interferes with the process of repression, when that inflection "gets in his mind" and disrupts the act of *imagining* wilderness.

This sort of repression is characteristic of an entire historiography of wilderness, a historiography to which McKibben clearly subscribes, most notably in his lengthy discussion of the writing of George Catlin. McKibben quotes selectively from Catlin's journal, editing into existence a vision of pure wilderness by cutting away any hint of "human society and of people"—in this case Native American society and people. In an act of the imagination comparable to his forgetting of the summer homes at his Adirondack lake, McKibben singles out a long passage, a description of a valley that was, according to Catlin, "far more beautiful than could be imagined by mortal man." In McKibben's words, it was one of those increasingly rare "visions of the world as it existed outside human history." This vision "sticks" in his "mind as a baseline, a reminder of where we began." It is a vision whose escape out of human history is an escape into myth, and the author makes no bones about its Edenic appeal: "If this passage had a little number at the start of each sentence, it could be Genesis" (52).

Of course, McKibben can imagine this valley as Eden only if he can imagine Catlin as Adam, and Catlin can only be Adam if he is the first human on the scene. McKibben must thus read Catlin rather perversely, erasing the presence of the many people who were not only omnipresent in the area but were the very reason Catlin had gone West to begin with. This Native American presence frames Catlin's depiction of the "wild" valley with a power and import that McKibben's editorial scalpel conveniently excises. Here is a sample of that elided narrative frame. Catlin writes: "The Indians, also, I found, had loved [this valley] once, and left it; for here and there were their solitary and deserted graves, which told, though briefly, of former chants and sports; and perhaps, of wars and deaths, that have once rung and echoed through this silent vale" (105). To be sure, Catlin himself minimizes the native presence in his Eden, if only by relegating it to the past. Describing in such glowing terms a landscape in which Native Americans have *only* a past, he constructs a landscape in which white Americans may have an uncontested future, a beautiful "prospect" in both senses of the term (Pratt 125). But Catlin cannot, in a book that is after all *about* Indians, banish that presence entirely; the completion of the job requires McKibben's additional editorial work.

For Catlin this effacement has just begun, so that the repressed native presence remains perilously close to the surface. The shallow Indian grave yet "speaks" to him, and the sound of all those chants, sports, and wars—all that prior humanity and culture, all that disruptive *history*—has clearly "gotten into his mind," as McKibben might say. Thus we are not at all surprised to find Catlin prefacing his description of wilderness-as-Eden by relating the following strange occurrence. He is camping out in the beautiful valley, peacefully, it seems, sleeping under the stars, when in "the middle of the night I waked, whilst I was lying on my back, and on half opening my eyes, I was instantly shocked to the soul, by the huge figure (as I thought) of an Indian, standing over me, and in the very instant of taking my scalp!" Catlin is momentarily "paralysed" by a "chill of horror" (103)—and then realizes he is looking up at his horse. One need not be Freud to see here the return of the repressed, of that which must inevitably be relegated to environmentalism's "political unconscious" in constructing the Edenic wilderness, in making an occupied territory into a scene of origin.

Reframing the Prospect
Identity? Nature = Postnaturality

> *There is no nature, only the effects of nature: denaturalization or naturalization.*
>
> JACQUES DERRIDA (QTD. IN BUTLER 1)

McKibben invokes Catlin's wilderness as if it were a "real" wilderness, an example of the sort of "first nature" that might yet ground our own imaginative constructions of wilderness and identity in the postnatural twentieth century. Yet clearly Catlin himself, in the supposedly more authentic nineteenth century, had *made* a wilderness, via the same sort of repression McKibben used to make a wilderness of his Adirondack lake. Thus do the groundings of ecological identity give way; the center does not hold. Examined closely enough, the putatively wild original turns out to be itself a copy, and I would argue that what in fact constitutes "postnaturality"—the condition of the subject after "the end of nature"—is this characteristically postmodern realization of the groundlessness of putatively natural identities. Postnaturality is not, as McKibben would have it, about the ubiquity of pollution or the global nature of climatic change. It is not about nature at all, but about fractures in the "natural" underpinnings of identity; it is the deconstruction of the ecologized subject.

We should expect postnaturality somehow to disrupt the disciplinary effects of ecologicality, though doubtless wilderness will continue, to return to Roderick Nash's phrase, to serve in some way as "a cultural and moral resource." Whether postnaturality will work to disentangle environmentalism from the retrograde political forces with which it has so long been intertwined remains to be seen. As ecocritics we can improve the odds, however, if our work places history once again before nature—if we follow Roland Barthes's advice "constantly to scour nature . . . in order to discover History there, and at last to establish Nature itself as historical" (101). Among other things this means approaching our national environmental narrative not, as Catlin and so many others have done, as the discovery and preservation of wilderness, but as the conquest of occupied territories and the recurring performance of new identities.

Notes

1. In *Strange Weather* and *The Chicago Gangster Theory of Life*, Andrew Ross examines the ways in which environmental discourse operates in the interest of a rapidly expanding global capitalism. His perspective is explicitly left and postmodern, and one of his favorite subjects is the profitable realm of green consumerism and its corporate counterpart, "greenwashing." Joni Seager, pursuing a similar project from an explicitly feminist perspective, explores in *Earth Follies* how environmentalism has helped to construct masculinity and legitimize patriarchy. For the environmentalist, the latter is perhaps the more sobering work; where Ross exposes the capitalist

underpinnings of ecotourism, The Nature Company, and the like—targets that seem peripheral to environmentalism as a political movement—Seager highlights the conservative gender ideology of even the most deeply committed and apparently progressive of green groups, most notably Earth First! and Greenpeace. The work of both critics has been less than warmly received by some in the ecocritical community; many would probably refuse to call it "ecocriticism" at all. I would argue that such a dismissal is misguided, however, if only because the target of Ross's and Seager's critique is not environmentalism per se—not the conscious effort to preserve a biologically diverse and livable world—but *ecologicality*, environmentalism as a discourse of subjectivity whose unintended effects can indeed be politically repugnant.

2. Slovic 4. This anthropocentric project of self-discovery appears at first to be circular: if such writers have little idea of their own boundaries prior to their encounters with external nature, how can they recognize nature as a "separate realm" in the first place? The truth, of course, is that they already have some sense of "who they are" when they begin their wilderness quests. In the early national period in which Catlin wrote, writers already knew themselves as distinct from nature initially by virtue of their race and their sex, and later by virtue of their nationality—as (primarily) white men and more generally as builders and representatives of a distinctively American "civilization." These senses deeply colored the construction of both the writers' own subjectivities and the environment they wished to preserve.

Works Cited

BARTHES, ROLAND. *Mythologies*. Tr. Annette Lavers. New York: Hill and Wang, 1995.

BUTLER, JUDITH. *Bodies that Matter: On the Discursive Limits of "Sex."* New York: Routledge, 1993.

CATLIN, GEORGE. *Letters and Notes on the Manners, Customs, and Condition of the North American Indians*. Vol. 1. London, 1841.

CRONON, WILLIAM, ED. *Uncommon Ground: Rethinking the Human Place in Nature*. New York: Norton, 1995.

GREENBLATT, STEPHEN. "Toward a Poetics of Culture." Ed. H. Aram Veeser. *The New Historicism*. New York: Routledge, 1989. 1–14.

MAZEL, DAVID. "Environmentalism as Domestic Orientalism." Ed. Cheryll Glotfelty and Harold Fromm. *The Ecocriticism Reader:*

Landmarks in Literary Ecology. Atlanta: U of Georgia P, 1996. 137–46.

MCKIBBEN, BILL. *The End of Nature*. New York: Random House, 1989.

NASH, RODERICK. *Wilderness and the American Mind*. New Haven: Yale UP, 1973.

OATES, JOYCE CAROL. "Against Nature." *Writing Nature: An Ecological Reader for Writers*. Ed. Carlyn Ross. New York: St. Martin's P, 1995. 460–68.

PEARCE, ROY HARVEY. *The Savages of America: A Study of the Indian and the Idea of Civilization*. Baltimore: Johns Hopkins UP, 1965.

PRATT, MARY LOUISE. "Scratches on the Face of the Country; or, What Mr. Barrow Saw in the Land of the Bushmen." *Critical Inquiry* 12 (Autumn 1985): 119–43.

ROSS, ANDREW. *The Chicago Gangster Theory of Life: Nature's Debt to Society*. London: Verso, 1994.

———. *Strange Weather: Culture, Science, and Technology in the Age of Limits*. London: Verso, 1991.

SEAGER, JONI. *Earth Follies: Coming to Feminist Terms with the Global Environmental Crisis*. New York: Routledge, 1993.

SLOVIC, SCOTT. *Seeking Awareness in American Nature Writing: Henry Thoreau, Annie Dillard, Edward Abbey, Wendell Berry, Barry Lopez*. Salt Lake City: U of Utah P, 1992.

SNYDER, GARY. *The Practice of the Wild*. San Francisco: North Point P, 1990.

THOMASHOW, MITCHELL. *Ecological Identity: Becoming a Reflective Environmentalist*. Cambridge, MA: MIT P, 1995.

Nathaniel Hawthorne Had a Farm
Artists, Laborers, and Landscapes in The Blithedale Romance

KELLY M. FLYNN

For narrator Miles Coverdale and the other inhabitants of Blithedale, as for Hawthorne himself at Brook Farm in West Roxbury in mid to late 1841, one desire overwhelms all others: to erase the distinction between what Nicholas Bromell has called mental and manual labor (12). The natural landscape, they feel, will provide inspiration to the poet and impetus to the farm laborer; instead of emphasizing the separate nature of these laborers, the members of the utopian community believe that "the twain shall meet" in each individual. The importance of this tenet to the real-life forerunners of the Blithedale inhabitants is clear in a letter written in 1840 by George Ripley, the founder of Brook Farm, to Emerson, asking for his support of the endeavor. Ripley enumerates the objectives of the community:

> to insure a more natural union between intellectual and manual labor than now exists [and] to combine the thinker and the worker, as far as possible, in the same individual . . . thought would preside over the operations of labor, and labor would contribute to the expansion of thought; we should have industry without drudgery, and true equality without its vulgarity. (qtd. in Sams 6)

The longed-for "union" between mental and manual is problematized, however, by the emergence of a subtly hierarchical discourse in which thought "presides over" labor and "equality" is matter-of-factly associated with an ominous "vulgarity."

The tendency apparent in this letter to use and to complicate the discourse of mental and manual labor is also visible in Hawthorne's letters, written during his seven-month sojourn at Brook Farm to his fiancée Sophia. An examination of Hawthorne's letters and of his journal entries from this period provides intriguing insights into *The Blithedale Romance*, begun a decade after his departure from West Roxbury. The letters, written to a private audience, concentrate on his laboring presence in the natural landscape and portray it as at once comic, tragic, and reinstitutive of the mental/manual dichotomy. The journal entries, in contrast, largely conceal that presence and focus instead upon the body of the land around him; in rarely mentioning his own manual labor, Hawthorne remains free—secure in his position as author—to assert his perspective on the landscape through his artistic labors. The novel serves as an attempt to grapple with these disparate narrative strategies; by creating a fictional account of an author laboring both physically and intellectually in a natural setting, Hawthorne experiments with immersing himself in the sweat and grit of the natural world while retaining an increasingly tenuous measure of artistic perspective and authority.

A letter written by Hawthorne to Sophia soon after his arrival at Brook Farm exemplifies the wry humor of many of his early letters. The humor serves as an outlet for Hawthorne's doubts about his physical abilities:

> This morning, I have done wonders. Before breakfast, I went out to the barn, and began to chop hay for the cattle; and with such "righteous vehemence" (as Mr. Ripley says) did I labor, that in the space of ten minutes, I broke the machine. . . . After breakfast, Mr. Ripley put a four-pronged instrument into my hands, which he gave me to understand was called a pitch-fork; and he and Mr. Farley being armed with similar weapons, we all three commenced a gallant attack upon a heap of manure. (qtd. in Sams 13–14)

Despite his implied ineptness, Hawthorne invokes the celebratory union between head and hand that Ripley had anticipated. Indeed, an almost religious sentiment lurks beneath the easy humor; Hawthorne imagines himself "purified" after his exertions, and he closes the letter with "I feel the original Adam reviving within me" (qtd. in Sams 14). The work of the hands has sanctified the work of the head. Hawthorne's excitement is fleeting, however; in his later letters, the marriage between head and hand vanishes and is replaced with a deep distrust of the circumstances that require such union. Instead of purifying and facilitating his authorial efforts, his labor at the farm has become suffocating and tedious. The dung heap becomes a "gold mine": "I think this present life of mine gives me an antipathy to pen and ink, even more than my Custom House experience did. . . . After a hard day's work in the gold mine, my soul obstinately refuses to be poured out on paper" (qtd. in Sams 21).

Hawthorne's rhetoric has shifted sharply from the language of the sacramental to the language of the saleable; his work at Brook Farm has brought his literary labor into threatening proximity to the realm of commodification, of "piles of money" and of "gold." Michael T. Gilmore, in tracing the spread of the commodity form in the mid-nineteenth century, notes: "As more and more of the physical world falls under the domination of the market, objects increasingly assume a kind of doubleness. They exist in their own right as material objects, but as commodities they come to possess a second significance or function as bearers of value. They can be used and consumed, and they can be exchanged for money" (15). Hawthorne's physical labor at Brook Farm appears as a struggle in a gold mine; the dung heaps and the furrows in the field exist as objects in a natural landscape, but Hawthorne's perspective as laborer "in" and "under" that landscape causes him to fear that both his "soul" and his "pen and ink" are equally susceptible to immersion in the marketplace. When he attempts to portray his own physical labor, he runs the risk of submitting the authorial self to burial in "the commodified earth" (qtd. in Sams 30).

Hawthorne's sense that close contact with the natural world through manual labor deadens both the "thinking" head and the "feeling" soul points to his Emersonian desire to integrate all the parts of the natural landscape; this desire is apparent in the contrast between the letters and the journals. The latter seem nothing so much as an exercise in regaining an intellectual and artistic perspective on Brook Farm's natural setting and on the labor that occurs within it. Hawthorne retires his "fine workman" persona by eradicating all traces of his own manual labor from the text (qtd. in Sams 15). A later commentator on the Brook Farm experiment, George William Curtis, writes tellingly of Hawthorne's impulse to remove himself from immediate involvement with the landscape: "Hawthorne was a sturdy and resolute man, and any heap of manure that he attacked must yield; but he had not come to Arcadia to sweat and blister his hands, and his blank and amused disappointment is evident. He had a subtle and pervasive humor, but no spirits. . . . He is always a spectator" (269).

The author was himself aware of this proclivity; early in the journal he remarks bluntly, "I, whose nature it is to be a mere spectator both of sport and serious business, lay under the tree and looked on" (*American* 202). Hawthorne's position as spectator is most telling in his accounts of the countryside around the community, accounts in which he consistently creates a distanced and intellectualized perspective on his surroundings. His language is that of the artist or the natural historian—he offers carefully sketched interpretations of the landscape. The land is, in the journal, the subject of the artist's gaze. Instead of the locus for physical labor, it becomes fodder for a literary imagination: "the earth looks more like a picture than anything else" (215).

Hawthorne is able to maintain this idealized vision of the natural landscape only as a function of distance. When he temporarily abandons his detached perspective and comes as close to the earth as he was forced to in his efforts in the "gold mine," the land becomes a disturbing agent of deception. The glorious hues mask an encroaching decay: "I have taken a long walk, this forenoon. . . . Most of this road lay through a growth of young oaks principally; they still retain their verdure; though, looking closely in among them, you perceive the broken sunshine falling on a few sear or dark-hued tufts of shrubbery" (205). A week later, Hawthorne writes, "None of the trees, scarcely, will now bear close examination; for then they look ragged, wilted, and of faded, frost-bitten hue; but at a distance, and in the mass, and enlivened by sunshine, the woods have still somewhat of the variegated splendor which distinguished them a week ago" (218); and reiterates in conclusion, "[The blueberry bushes'] hue, at a distance, is a lustrous scarlet; although it does not look nearly so bright and beautiful, when examined close at hand. But, at a proper distance, it is a beautiful fringe on Autumn's petticoat" (220). Hawthorne the writer here becomes Hawthorne the imagist.

In his avowal that the landscape's beauty depends upon his distance from it, Hawthorne echoes the long-standing Romantic tradition of landscape painting. The ability of the leisured class to experience the American landscape was in the nineteenth century directly predicated upon the initial presence of manual labor. It is surely not a coincidence that the emergence of the Hudson River School—devoted to portraying that "wild, romantic, and awful scenery" that the travelers discovered—can be dated from 1825, the same year that the Erie Canal was completed (Hunt 109). Without the canal-digger's efforts, artists such as Thomas Cole might not have been able to influence and to capitalize upon a growing interest in the coloring and outline of the American landscape. Interestingly, in his "Essay on American Scenery" (1836), Cole anticipates somewhat ambiguously the increasing presence of both manual and mental laborers in that landscape: "Where the wolf roams, the plough shall glisten; on the gray crag shall rise temple and tower—mighty deeds shall be done in the now pathless wilderness; and poets yet unborn shall sanctify the soul" (8). Does the land act as a unifying force, bringing together and setting equal worth on the endeavors of poet and ploughsman? Or does it serve as a catalyst for the extension of the mental/manual dichotomy by forcing the artist to reassert his impact on the land in the face of the admittedly "mighty deeds" of the physical laborers?

The literary "eye" is, for Hawthorne, the best means of assimilating and "entering" the natural landscape. In contrast, his bodily presence in the environment is most often figured as obtrusive and somehow improper. He refers to himself as an "intruder" when he enters the cattle pasture—the animals look at him with "long and wary observation." The pine woods on the edge of the pasture resist his infiltrating presence: "The whole is tangled and wild, thick-set, so that it is necessary to part the rustling stems and branches, and go crashing through" (197). When he walks through the woods near Pulpit Rock, "a crow is aware of [his] approach a great way off, and gives the alarm to his comrade loud and eagerly" (214), and he seems to regret lying among oaks and white-pines "so close together that their branches intermingled" and watching a squirrel: "The squirrel seemed not to approve of my presence; for he frequently uttered a sharp, quick, angry noise, like that of a scissor-grinder's wheel" (215). In fact, whenever Hawthorne abandons his authorial subjectivity and immerses himself in the landscape, collapsing the distance between self and landscape, he risks becoming the object of others' interpretations and observations. He must cling tenuously to his final affirmation of the power of the mental laborer to embrace and possess the natural world. The rhetorical moves to metaphorize the actual landscape in the journals bring the earth safely into the distanced domain of the writer. The land is his to depict, but he is not its to disturb and suffocate.

The Blithedale Romance is the testing ground for the conflicting impulses expressed in Hawthorne's letters and journals, i.e., the need to express oneself through manual labor and the need to assert one's authorial gaze over that labor. Narrator Miles Coverdale leaves Boston so that he may work in a truly natural environment; he will seek to redeem his soul, and his literary worth, in an outdoor setting free from artificiality and degeneration. But just as the farmhouse contains within it social strictures that the utopians had hoped to escape, so too do the fields reflect the competitiveness of a commercial culture: "It struck me as rather odd, that one of the first questions raised, after our separation from the greedy, struggling, self-seeking world, should relate to the possibility of getting the advantage over the outside barbarians, in their own field of labor" (*Blithedale* 20). Coverdale finds that his worth will be measured by his efficiency in this competition as Silas Foster notes dryly that "three of you city-folks [are] worth one common field-hand" (20). The narrator initially refutes his implied incompetence by reinventing the dichotomy between mental and manual. The labor of the head is no longer to be set above the labor of the hand, and the work of the literary intellectual is figured as "enervating" and "weary" (19). As Hawthorne begins to explore the place of his fictional author in the landscape he found personally oppressive, Coverdale's problematic relationship to manual labor reemerges through his growing concern with the status of his own and others' bodies within the body of the land itself. Just as the "gold-mine" threatened to suffocate Hawthorne in his letter to Sophia, in *The Blithedale Romance* the immersion of the self in "the enlightened culture of the soil" threatens to destroy one's ability to see oneself in accurate relation to the physical world (61). For Coverdale and for Hawthorne, vision is inextricably associated with the authorial identity. The tendency of physical labor to cloud that vision is framed as a judgment on the hand that farms and writes: Coverdale's manual labor both interferes with his literary authority and violates his selfhood. His attempts to participate in the natural landscape have somehow "backfired" and his position becomes analogous to Hawthorne's own position in the journals. When Hawthorne entered the natural landscape around Brook Farm, he risked effacing his own subjectivity and assuming the status of object in another's gaze. Coverdale is prey to a similar anxiety. His labors at the farm endanger his status of author. Rather than manipulating the labor of others in a literary context, he is himself converted into grist for a storytelling mill. He catalogues the "slanderous fables" told by the community's neighbors about the poet-farmers' comic ineptitude in plowing, milking, and gardening, and concludes indignantly: "Finally, and as an ultimate catastrophe, these mendacious rogues circulated a report that we Communitarians were exterminated, to the last man, by severing ourselves asunder with the sweep of our own scythes!" (64–65).

The objectification of his own involvement with the earth is indeed the "ultimate catastrophe" for Coverdale. His anger at the "mendacious rogues" is akin to Thoreau's outrage at the "unclean and stupid farmer" who had the audacity to name a pond near his cabin in the Walden woods (195). In each case, the hands have assumed a duty "sacred" to the head, and the appropriation of the authorial function is perceived as a threat. Later in the text, Coverdale reasserts his own authorial presence by offering to Hollingsworth an alternate "history" of Blithedale and of his (Coverdale's) simultaneously mental and manual position in that history. By figuring himself in the prediction as the best of physical laborers, and as the teller of that prediction in the larger narrative, Coverdale combines Hawthorne's own positions of immediate involvement in the natural landscape, in his letters, and of removed interpretation of that landscape, in his journals. Indeed, he echoes Hawthorne's relation to *The Blithedale Romance* by "objectifying" himself in an imagined "Epic Poem."

The narrator's projection of himself into a literary future is one of three attempts to assert his authorial "eye." He also combats the deprivileging of his authoritative gaze by appropriating the vision of others and by reinscribing his own perspective on the natural landscape. The second of the three strategies is clearest when old Moodie arrives at Blithedale in search of the frail and mysterious Priscilla, thus stimulating the narrator's interest and sympathy: "I tried to identify my mind with the old fellow's, and take his view of the world, as if looking through a smoke-blackened glass at the sun. It robbed the landscape of all its life." The "pleasantly swelling slopes," the "peculiar picturesqueness of the scene," the "twinkling showers of light" that fall into the forest, are immediately infused with "the cold and lifeless tint of [Moodie's] perception" (84). It remains for Coverdale to diffuse the power of the landscape by isolating himself from its connection to manual labor and by reinterpreting its effect as artistic and literary. If he does not, he is left to a view of the natural world and of his position in it as depressing as the one he saw through Moodie's eyes. Far from being the "Paradise" from which springs the renewed Adam (16), the natural environment is instead the breeding ground of the fall into polarization between mental and manual, between poetic and physical: "The yeoman and the scholar are two distinct individuals, and can never be welded into one substance" (66). Coverdale's final attempt to assert his unchallenged subjectivity must lessen the danger of immediate contact with the landscape.

Hence, almost immediately after his meeting with Moodie, Coverdale begins to characterize his relationship to nature in an entirely new manner. He retreats from his propensity for human contact—which "impelled" him to "live in other lives"—and writes of his need for solitude (160). The landscape is refigured as a source of "drowsy pleasure" and a means of removing "the ache of too-constant labor from [his] bones" (89). The countryside functions solely as a mirror of his inner emotion: "Everything was suddenly faded. The sun-burnt and arid aspect of our woods and pastures, beneath the August sky, . . . symbolize[d] the lack of dew and moisture that since yesterday, as it were, had blighted my fields of thought, and penetrated to the innermost and shadiest of my contemplative recesses" (138). In making the natural landscape an extension of the self, Coverdale reinscribes his own perspective upon it; he shifts from an exploration of others' interpretations of the land to a celebration of the artist's relation to it. The iconography of his "pine bower" reflects most fully his reassertion of an authority more mental than manual. He describes his dwelling as

> a kind of leafy cage, high upward in the air, among the midmost branches of a white-pine tree. A wild grapevine, of unusual size and luxuriance, had twined and twisted itself up into the tree. . . . This hermitage was my one exclusive possession, while I counted myself a brother of the socialists. It symbolized my individuality, and aided me in keeping it inviolate. (98–99)

Where Hawthorne "lay under a tree and looked on" (*American* 202), Coverdale perches in a tree and indulges his even more voyeuristic impulses. From this vantage, his literary eye is unchallenged. The grapevine is not a carefully cultivated crop: its fruit will make not wine (which carries with it the threat of reinsertion into a physical laborer's perpective), but allegory (which preserves and symbolizes the author's impact upon the environment): "It gladdened me to anticipate the surprise of the Community, when, like an allegorical figure of rich October, I should make my appearance, with shoulders bent beneath the burthen of ripe grapes, and some of the crushed ones crimsoning my brow as with a blood-stain" (99). Even the syntax of the passage acts to remove the hermitage from the sphere of manual labor, as active verbs and reflexive pronouns assert the "independence" of the vine. Coverdale's vision of nature has turned away entirely from his disturbing—for Hawthorne, suffocating—connection with it via "the virtue yet in the hoe and the spade": the land now functions solely as a source of literary inspiration (Emerson, "American" 62).

The pine bower neatly metaphorizes both Coverdale's own shift in emphasis from manual labor to mental fancy and the narrative's shift in depicting the West Roxbury landscape: no longer the milieu for "the curse of Adam's posterity," the land is now an arena for purely imaginative activity (206). The authorial construction and use of the land has reached its fullest potential. Only a decade after Hawthorne had initially proposed immersion in the natural world, he has retreated from its threatening implications, finding safety instead in an authorial strategy that echoes a sentiment expressed by Emerson in *Nature*: "There is a property in the horizon which no man has but he whose eye can integrate all the parts, that is, the poet" (9). The land can never be appropriated by manual labor; it can only be rendered comprehensible by the intellectual labor of the author.

Works Cited

BROMELL, NICHOLAS K. *By the Sweat of the Brow: Literature and Labor in Antebellum America*. Chicago: U of Chicago P, 1993.

COLE, THOMAS. "Essay on American Scenery." *American Monthly Magazine* N.S. 1 (1836). 6–15.

CURTIS, GEORGE WILLIAM. "Editor's Easy Chair." *Harper's New Monthly Magazine* 38 (January 1869). 268–71. Rpt. as "Hawthorne, Brook Farm, and Transcendentalism" in *The Brook Farm Book: A Collection of First-Hand Accounts of the Community*. Ed. Joel Myerson. New York: Garland, 1987. 92–100.

EMERSON, RALPH WALDO. "The American Scholar." 1837. Ed. Robert Spiller. Vol. 1 of *The Collected Works of Ralph Waldo Emerson*. Gen. ed. Alfred R. Ferguson. Cambridge: Harvard UP, 1971. 49–70.

———. *Nature*. 1836. Ed. Robert Spiller. Vol. 1 of *The Collected Works of Ralph Waldo Emerson*. Gen. ed. Alfred R. Ferguson. Cambridge: Harvard UP, 1971. 3–48.

GILMORE, MICHAEL T. *American Romanticism and the Marketplace*. Chicago: U of Chicago P, 1985.

HAWTHORNE, NATHANIEL. *The American Notebooks*. 1868. Ed. Claude M. Simpson. Vol. 8 of *The Centenary Edition of the Works of Nathaniel Hawthorne*. Gen. Eds. William Charvat, Roy Harvey Pearce, and Claude M. Simpson. Cleveland: U of Ohio P, 1972.

———. *The Blithedale Romance*. 1852. Ed. Fredson Bowers. Vol. 3 of *The Centenary Edition of the Works of Nathaniel Hawthorne*. Gen. eds. William Charvat, Roy Harvey Pearce, and Claude M. Simpson. Cleveland: U of Ohio P, 1972.

HUNT, FREEMAN. *Letters About the Hudson River*. New York: Freeman Hunt and Co., 1836.

SAMS, HENRY W. *Autobiography of Brook Farm*. Englewood Cliffs, NJ: Prentice-Hall, Inc., 1958.

THOREAU, HENRY DAVID. *Walden*. 1854. Ed. J. Lyndon Shanley. Princeton: Princeton UP, 1971.

Agrarian Environmental Models in Emerson's "Farming"
STEPHANIE SARVER

Late in 1858, Ralph Waldo Emerson delivered an address to the Middlesex Cattle Society that became known as "Farming"; it was later published in *Society and Solitude* (1870). In the address, Emerson brings his philosophy of nature to an agricultural milieu to suggest how farmers may realize spiritual rewards through their work. He approaches farming as a topic of philosophical inquiry. Rather than attempt to fully characterize how the farmer relates to nature, Emerson explores three seemingly incompatible models for explaining that relationship: he suggests that the farmer exists within a natural order to which he must adapt; that the farmer dominates agrarian nature through his knowledge of science; and, that the farmer is the minder of an agrarian factory created by God. Emerson's effort to describe how the farmer relates to nonhuman nature resonates with twentieth-century attempts to identify an environmental ethic for farmers, who are subject to the forces of nature as they simultaneously work to control them.[1] Emerson's use of perceptual models to describe farmers' interactions with nature help explain his difficulty in drawing on farming as an activity that might promote human spirituality.

"Farming" represents Emerson's attempt to apply an individual experience of nature to the social and economic activities practiced in agriculture. "Farming" figures as a synecdoche of Emerson's world. Its rambling structure carries us through several models for explaining the relationship between the farmer and nature, and suggests a rhetorical resistance to monolithic models for understanding how humans interact with nature. Emerson's earlier work suggested that transcendence is achieved in solitary connection with nature, but in "Farming" he suggests that agriculture may also provide opportunities for such a connection. Throughout the address, however, he further reveals that agriculture is implicated in the larger community and marketplace. Because farming is enmeshed with society, its potential to figure as a spiritual practice is always at risk since it exists between the often conflicting forces of unmediated nature and civilization.

In the opening paragraph of "Farming," Emerson reveals that he will consider nature in its common and philosophical dimensions, both of which he explored in *Nature* (1836).[2] He acknowledges the ambiguity surrounding the concept of nature by distinguishing the terrestrial realm, or "the earth," from a larger inclusive and amorphous "Nature." The farmer, he suggests, exists "close to Nature" and "obtains from the earth the bread and the meat" (*Complete* 7:137). This sentence figures as the introduction to a discussion of the farm, which, in its common definition, is that realm wherein the farmer transforms nature. Defined philosophically, the farm is an expression of a larger amorphous nature, which may be both the reflection of the farmer and the medium through which the farmer may know and understand God and transcend material existence. These several meanings accrete through the address as Emerson examines how farming may be an activity of spiritual utility.

Emerson initially describes the farmer's relationship to nature by suggesting that the farmer is a student who adapts to natural conditions when he "bends to the order of the seasons" (*Complete* 7:138). The farmer practices a habit of patience that engenders a reverence and respect for nature. Responding to the larger natural world, he understands that "the earth shall feed and clothe him; and he must wait for his crop to grow" (*Complete* 7:139). This adaptation binds the farmer to the earth. He "clings to his land as the rocks do," which implies that he is the subject of environmental forces (*Complete* 7:139). In suggesting that the farmer figures as a part of a larger natural network, Emerson invokes what we may interpret as a proto-ecocentric egalitarian view when he identifies humans as a species inherently no more valuable than any other species within a given natural community. Such a view entails a "deep-seated respect, or even veneration, for ways and forms of life" (Naess 95). Indeed, while

Emerson's view is arguably "ecological" in our modern sense, he nonetheless contends that farmers are sensitive to the ways of nature and that they recognize the interconnectedness of all of creation.

Although Emerson begins the address by suggesting that the farmer works in response and adapts to nature, he soon acknowledges that this only partially describes farming, which also entails manipulating nature. Thus, Emerson shifts from a model in which the farmer is a passive figure to a model depicting a mechanistic relationship between the farmer and nature. He moves from suggesting that the farmer merely "clings to his land" to suggesting that nature is a perfectly timed machine minded by the farmer. The farmer's task is to understand the machine; the tools that will facilitate his understanding are the sciences. Emerson tells us: "the earth is a machine. . . . Every plant is a manufacturer of soil. . . . The plant is all suction-pipe,—imbibing from the ground by its root, from the air by its leaves, with all its might" (*Complete* 7:144). By shifting to a mechanistic model, Emerson readjusts the view that the farmer is merely one entity subject to natural forces; instead, he suggests, the farmer can study nature and thereby manage and manipulate it.

In developing a model of the farmer as both the passive object of nature and the minder of its mechanism, Emerson reveals how models for perceiving human relationships with nature may be caught between seemingly contradictory forces. The farmer, Emerson claims, is "timed to Nature, and not to city watches" (*Complete* 7:138). By juxtaposing nature with machines, he directs us to consider how the farmer is situated within an agricultural economic system that operates increasingly on factory models. The farmer, who is governed by the seasons, is nonetheless caught within an economy that operates like a machine, and which is dependent upon a mechanized conception of time.[3] In addressing the "city watches," Emerson obliquely speaks to the way that industry has reduced the "great circles" of nature to ever smaller increments. The farmer is necessarily controlled by nature's timing, but he is also increasingly the subject of the city clocks of market forces.

As he advances through his address, Emerson attempts to reshape the image of the farmer as a passive figure by investing him with not only power over a mechanistic nature, but also power over society. In doing this Emerson invokes a third model of stewardship in which humans figure as mediating agents who stand between God and nature. In this model, the anthropocentric farmer tends and nurtures nature according to a divine design and is rewarded for his effort. Emerson's model of stewardship acknowledges that the farmer does not exist in isolation from society, and that his activities are not directed solely at attaining a spiritual connection to nature. He observes that the farmer produces the food upon which a society sustains itself, thereby mediating between nature and society rather than between nature and God. Thus, he tells us, "In the great household of Nature, the farmer stands at the door of the bread-room, and weighs to each his loaf" (*Complete* 7:140). The farmer manages nature and the farm, which in turn reward him for his wise and sensitive care.

The farmer not only experiences nature in a way that benefits him individually, he also manipulates nature for the good of his country. Emerson tells us, "He who digs a well . . . plants a grove of trees by the roadside . . . makes a fortune which he cannot carry away with him, but which is useful to his country long afterwards" (*Complete* 7:141), alluding to the extent to which the farmer is implicated in a larger social complex: nature influences the individual farmer and, by extension, entire nations. Emerson suggests that the farmer's understanding of nature ultimately figures in the integrity of the larger social structure. As either a minder or a steward, however, Emerson's farmer has a limited ability to enjoy an intimate experience of nature; in one case he manages nature, and in the other he mediates between nature and society, always enjoying a close proximity to a natural realm yet never existing within it as he would when working as a solitary farmer.

It is noteworthy that Emerson's ideal farmer is engaged in a solitary activity. Emerson initially identifies his farmer as the "first man," a twist on Jefferson's more pluralistic view of farmers as the chosen people of God. Emerson draws upon Biblical imagery traditionally associated with American agrarianism to identify farmers as the cultivators of a promised land, but he places his farmer in an ancient garden as the first man to whom all men look with respect. When Jefferson suggests that farmers are the chosen people, he invokes the image of a community engaged in a divine enterprise; conversely, when Emerson invokes the Adamic model, he suggests that the farmer functions in a solitary role, and that his activities are centered in the individual rather than in the larger community.

Emerson's emphasis on the relationship between the individual and nature represents an adaptation of the theories he develops in his earliest work. In "The Uses of Natural History" (1833), he suggests that the farmer is in a position to realize a serenity in his experience of nature. He says, "Dig your garden, cross your cattle, graft your trees, feed your silkworms, set your hives—in the field is the perfection of the senses to be found, and quiet restoring Sleep" (*Early* 11). This experience requires that the farmer be autonomous; the quiet restoring sleep of the fields will likely be found only when the farmer enjoys an unfettered communion with agrarian nature. Emerson's allusion to breeding cattle and grafting trees, however, also speaks to the incursion of society on the farmer's activities.

While the farmer may enjoy spiritual benefits from his communion with nature, he nonetheless works in a world that, even in 1833, was influenced by the market and by scientific approaches to farming. The agriculture to which Emerson alludes in "The Uses of Natural History" figures in not only a spiritual, but also a material economy in which the larger community is implicated.[4] Emerson resists involving himself in this reality even as he acknowledges it, rhetorically working to preserve the farm as a site for spiritual inquiry rather than social reform.[5] In "Farming," Emerson sidesteps complications posed by the social and political implications of farming by directing his focus to the solitary individual: the farmer.

As in many of his essays, in "Farming" Emerson does not follow a straight course through a set of logical arguments to reach a clear conclusion. Instead, he leads us through several views of how farmers might perceive their relationships with both nature and society toward the end of realizing an experience that emphasizes a spiritual apprehension of the agrarian cosmos. He addresses the farm, the farmer, and the activity of farming, describing how one can operate simultaneously within actual and metaphoric realms. Emerson describes the farmer as a symbol of nature (Corrington 20), but he also reveals that the farmer "stands close to Nature," and that he knows and manipulates a complex of natural conditions (*Complete* 7:137). Emerson shifts from depicting the farmer as adapting to nature ("He bends to the order of the seasons" [*Complete* 7:138]) to depicting the farmer as mastering nature: "The earth works for him; the earth is a machine which yields almost gratuitous service to every application of the intellect" (*Complete* 7:144).

When we consider the various views of nature with which Emerson works, agriculture occupies an ambiguous position. How should farmers perceive their relationship to nature? Emerson tells us that nature is no more than "essences unchanged by man," which suggests that the farm may well figure as an unnatural realm (*Complete* 1:5); this may explain Emerson's comfort in describing the farm as a kind of machine and the farmer's relationship with it as mechanistic. However, Emerson also suggests that farming may be a "natural" activity; therefore, it could figure as an outward manifestation of an inward experience, a representation of the signs of "natural facts" upon which farmers might draw in their own spiritual improvement.

This pervasive ambiguity about where farming fits in Emerson's philosophy points to the problem of working with models for perceiving human relationships with nature, and the related difficulty of classifying certain human activities, such as farming, as inherently ecocentric or anthropocentric. As we have seen in "Farming," Emerson shifts among various models to explain the relationship between the farmer and nature. He first presents a hierarchical model of the universe that positions nature at its apex, governing the farmer; he then slightly revises that model to suggest that the farmer is the confidante of nature, understanding its secrets, but nonetheless still its subject. Finally, Emerson suggests that the farmer mediates between nature and society when he says "It is for him to say whether men shall marry or not" (*Complete* 7:140).

Emerson argues that the farmer is privileged by his relationship with and knowledge of nature. This sanguine assertion is actually a bit of rhetorical boosterism which directs us beyond the text to consider Emerson's context. Rather than being privileged, the New England farmer of 1858 was in fact experiencing an ever-dwindling influence as the manager of the nation's "bread-room." By the mid-nineteenth century, large-scale midwestern farms had forced New England farmers to become highly specialized in their food production. Grain was shipped by canals and railroads from western farms to eastern markets. New England, which had traditionally resisted scientific approaches to farming, faced increased pressures to adopt the more efficient agricultural techniques of western farmers.[6] Within this context, Emerson could not credibly promote his idealized view of farmers as existing only as isolated figures within the balance of nature; thus, through the course of his address he shifts to other modes to suggest that the farmer has control of an agrarian machine, and in doing so acknowledges the extra-textual realities of his address. Emerson suggests that the earth is a great factory and points to the ways in which the farm is a natural version of New England industry. Emerson's farmer is the manager of a machine of colossal proportions. He reminds his audience that the farmer's understanding of the machine of nature will impart spiritual rewards. Although he tries to emphasize the spiritual benefits of farming, the relationship between the farmer and nature is strained under the mechanistic images, which drag the individual farmer ever farther from essences unchanged by man and into the realm of New England factories.

Emerson attempts to contain the farm within the realm of nature but ultimately cannot extract it from the influences of civilization. The farmer, he suggests, is a "hoarded capital of health, as the farm is the capital of wealth" (*Complete* 7:140). As the capital of health, the farmer is a part of a spiritual economy in which his habit of mind and his labor yield spiritual rewards. But perhaps more important to Emerson's audience is that the New England farm represents a capital of wealth. In drawing this parallel, Emerson seems to demonstrate the impossibility of excising farming from the social realm.

It seems that Emerson hoped to simplify farming by focusing on the farmer as a solitary figure rather than a member of a larger community. By adapting the farmer to his already well-developed model of spiritual individualism, Emerson defies a more popular image of the farmer as a part of a larger community that forms the foundation of democratic society. His view is not incompatible with a Jeffersonian model of the farmer; but it is highly idealized and extracted from the actual context of the nineteenth-century world. In "Farming" Emerson repeatedly steps into the realm of politics, then withdraws. This move may be interpreted as redirecting our attention from the insubstantial concerns of civilization to the more important spiritual concerns of nature.

Some environmentalists have argued that the resolution of environmental problems will likely depend on developing solutions that acknowledge the dynamic interactions between the individual, the society, and the environment. Although Emerson was unconcerned with the state of the environment per se, his work nonetheless focuses on the relationships and dynamic interactions among these entities. He is concerned with nurturing the human spirit and looks to an intimate understanding of nature to accomplish this. Emerson invokes a mechanistic model of nature to explain the work of the farmer, yet he simultaneously acknowledges the limitation of that model by suggesting the extent to which the farmer is subject to nature. The farmer exists at the nexus of nature and culture; in nineteenth-century New England, that culture was increasingly mechanized. In shifting from ecocentric stewardship to mechanistic models to describe the relationship between the farmer and nature, Emerson demonstrates how agriculture may belong to both nature and human culture.

Emerson spoke to an agricultural world that was radically different from that in which we live today, yet his views of the relationship between the farmer and nature might be considered within the context of modern debates about agriculture and environmental ethics. Many environmentalists justly criticize modern agriculture for its reliance on techniques that maximize production at the expense of the diverse plants, animals, and insects that inhabit agricultural lands; their criticisms emerge from the agricultural practices of the past fifty years (Thompson 21–46). Many call for a revision of the way we perceive our relationship with nature and variously call for the adoption of ecocentric or stewardship models as necessary steps in mitigating current environmental problems and subverting future catastrophe.

Through its rhetorical permutations, Emerson's "Farming" demonstrates the challenge of identifying models for human practice that are sensitive to the subtleties of nature while still acknowledging the influences of a social complex that is often at odds with environmentally sensitive practices. In refusing to adhere to a single conceptual model, Emerson insists that agriculture is philosophically complex. "Farming" may offer a conceptual model of agriculture that allows for many conditions to exist simultaneously: Emerson's model for farming acknowledges that the farmer may feel a reverence for nature even as he exercises control over it. He also suggests that the farmer has an obligation and responsibility to the human society that relies upon his crops for its well-being and yet that the farmer ultimately remains simply one entity among many within the larger natural cosmos.

Notes

1. For example, see the work of Liberty Hyde Bailey, Aldo Leopold, and, more recently, J. Baird Callicott, Wes Jackson, and Wendell Berry.
2. Common nature is composed of "essences unchanged by man; space, the air, the river, the leaf" (*Complete* 1:5). The philosophical dimension of nature is more inclusive: it is "all that is separate from

us . . . that which is both nature and art, all other men and my own body" (*Complete* 1:4–5).

3. Richard Brown observes that the mass production of clocks increased rapidly after 1830, which made timepieces available to a wide population with the consequence that concepts of both time and industrial efficiency were transformed (134–5).

4. Michael T. Gilmore addresses this dual phenomenon in a similar way. Under the influence of the market, he suggests objects acquire a dual status as both material realities and as commodities with exchange value. Emerson, as a symbolist, "perceives the world as twofold: things have a concrete reality . . . and they signify an idea or another object . . . 'exchangeable' for something else" (15).

5. Emerson was familiar with the farm as an instrument of social reform through his peripheral association with Brook Farm. He declined to join the commune, fearing that a communal venture would subordinate the individual spiritual experience of farming to a larger political agenda. Emerson's journal explains his resistance: "to join this body would be to reverse all my long trumpeted theory . . . that a man is stronger than a city, that his solitude is more prevalent and beneficent than the concert of crowds" (*Journals* 408). Gay Wilson Allen discussed Emerson's reluctance to join the commune; Emerson wrote to Brook Farm's founder, George Ripley, explaining, "I am . . . not as favorably disposed to his Community of 10 or 12 families as to a more private reform" (Allen 364).

6. For a discussion of the way that New England agriculture evolved in response to competition with farm products produced in the West, see Cochrane and Danhof.

Works Cited

ALLEN, GAY WILSON. *Waldo Emerson: A Biography*. New York: Viking P, 1981.

BROWN, RICHARD D. *Modernization: The Transformation of American Life, 1600–1865*. New York: Hill and Wang, 1976.

COCHRANE, WILLARD W. *The Development of American Agriculture: A Historical Analysis*. 2d ed. Minneapolis: U of Minnesota P, 1993.

CORRINGTON, ROGER. "Emerson and the Agricultural Midworld." *Agriculture and Human Values* 7:1 (Winter 1990): 20–26.

DANHOF, CLARENCE H. *Change in Agriculture: The Northern United States, 1820–1970*. Cambridge: Harvard UP, 1969.

EMERSON, RALPH WALDO. *The Complete Works of Ralph Waldo Emerson*. 1903–4. New York: AMS P, 1979.

———. *The Early Lectures of Ralph Waldo Emerson*. Ed. Stephen E. Whicher and Robert E. Spiller. Vol. 1. Cambridge: Harvard UP, 1959.

———. *The Journals and Miscellaneous Notebooks of Ralph Waldo Emerson*. Ed. A. W. Plumstead and Harrison Hayford. Vol. 7. Cambridge: Belknap P of Harvard UP, 1969.

GILMORE, MICHAEL T. *American Romanticism and the Marketplace*. Chicago: U of Chicago P, 1985.

NAESS, ARNE. "The Shallow and the Deep, Long-Range Ecology Movement. A Summary." *Inquiry* 16: 95–100.

THOMPSON, PAUL B. *The Spirit of the Soil*. London: Routledge, 1995.

Exploring the Linguistic Wilderness of The Maine Woods
ANN E. LUNDBERG

In concluding the "Chesuncook" section of *The Maine Woods*, Thoreau reflects that "The partially cultivated country it is which chiefly has inspired, and will continue to inspire, the strains of poets, such as compose the mass of any literature." The lines draw us back toward Walden, itself the partially cultivated country between Concord and the West, and the place from whence his own most celebrated poetical work would spring. The passage goes on to remind us, however, that "not only for strength, but for beauty, the poet must, from time to time, travel the logger's path and the Indian's trail, to drink at some new and more bracing fountain of the Muses, far in the recesses of the wilderness" (155–56). If in *Walden* Thoreau momentarily manages to bridge the gap between the written word and the natural world by means of poetic cultivation, in *The Maine Woods* he is intent upon exploring the wilderness remaining between the word and the world it is meant to represent. Where the textual model of *Walden* provides a sense of permanence, of objective existence inherent in its written words, the prevalence of the oral in *The Maine Woods* implies that the relation between word and thing is transient at best. Moreover, it suggests the possibility of silence, of absences within language. This implication in turn affects Thoreau's treatment of the written word.[1] The logger's path and the Indian's trail represent the contrasting and to some degree competing modes by which Thoreau enters the woods in pursuit of a true language.

Where the logger's way is the way of the written word, of the map, of the writer, the Indian's is the way of experience, of an oral mode of being. Rather than subsuming the oral to facilitate writing, as he does in *Walden*, in *The Maine Woods* Thoreau encounters the spoken word on new ground.

For Thoreau, as for other authors working within the genre, nature writing involves a process of mapping by which he attempts to place himself more surely in the natural world. This need for self-location also suggests the possibility of being lost, of not being at home in the woods or in the words used to describe them. When such disorientation does occur, nature writing exhibits an anxiety which reduplicates the experience of the foreignness of the woods and the lack of understanding within the self.[2] For Thoreau the woods speak of potential presences concealed in the shades of the forest: the wandering moose, the snake in the leaves, the falling tree, all of which are heard but remain unseen. The Maine woods are also the site of Thoreau's encounter with the language and ways of his Indian guides whom, for all the helpful direction they offer, he cannot fully comprehend. The combination of sound and silences implies the existence of something not available as legible text or as the object of visual possession. The language of the woods cannot be translated onto the page and thus reveals an absence both within the interpreting self and within Thoreau's working paradigm of language.

In order to understand the unsettling effect of the Maine woods upon Thoreau's ideas regarding language, it is useful to approach those woods by way of Walden, as did Thoreau. *Walden* cautions that by reading "only particular written languages, which are themselves but dialects and provincial, we are in danger of forgetting the language which all things and events speak without metaphor, which alone is copious and standard" (111). As long as we are mere reader, our vision is confined by the page. We should instead be broad-visioned Seers, who not only see but hear. Nature is not a passive text, but speaks, and in so doing, actively translates its meaning to human understanding. The question for the writer then becomes how to overcome the "memorable interval between the spoken and the written language, the language heard and the language read" (101).

Thoreau answers this question by creating the linguistic density of *Walden*, a density which derives from the gravitational effect of the pond, the concentrated center of meaning. Thoreau explicitly pronounces that linguistic coherence is achieved through the purifying process of writing. The language heard "is commonly transitory, a sound, a tongue, a dialect merely, almost brutish, and we learn it unconsciously, like the brutes, of our mothers. The [language read] is the maturity and experience of that; if that is our mother tongue, this is our father tongue, a reserved and select expression, too significant to be heard by the ear, which we must be born again in order to speak" (101). It is the written language, finally, which most clearly exhibits reality in its stable form: "the noblest written words are commonly as far behind or above the fleeting spoken language as the firmament with its stars is behind the clouds" (102). Selective revision, such as that which produced *Walden*, makes writing visionary. Whereas transitory speech does not achieve the word's full potential, writing makes the statement as expressive as the natural object, a mirror as reflective as the unruffled surface of Walden pond.

Thoreau's "ground" is the relation between the natural world and the mother tongue which the written word translates phonetically. The instability of Thoreau's linguistic ground derives from his ambivalence toward the mother tongue, which is the origin of the written word and yet is disturbingly transient. Cut off from its roots, the written word risks becoming a dead word, one separated from the natural world. What must be recovered is a genetic relation between nature as fact and nature as word.

That nature speaks becomes apparent in the "Spring" chapter of *Walden*, wherein Thoreau dispels his anxieties about the mother tongue. As Michael West has shown, Thoreau drew heavily upon Charles Kraitsir's linguistic theories for the chapter's triumphant conclusion. Kraitsir argued that the principles by which language is constructed are inherent in human nature and that language is objective, universal, and governed by discernible laws: "All men, however diverse they may become by conflicting passions and interests, have yet the same reason, and the same organs of speech. All men, however distant in place, are yet plunged into a material universe, which makes impressions of an analogous character, upon great masses" (1). Kraitsir theorized that, given the limits of physiology, the essential character of human nature, and the objective reality of the natural world, language cannot fail to signify concretely. Similarly, for Thoreau, the melting bank of *Walden*'s famous "sand foliage" passage provides an organic model of language wherein purely material nature, the "inert bank," flowers into significant form and simultaneously into human language. The flow, however, is momentary, contained, and finally reified. Matter flows into life much as a seed develops. Thoreau remarks, "No wonder that the earth expresses itself outwardly in leaves, it so labors with the idea inwardly. The atoms have already learned this law and are pregnant by it" (306). The creation of order out of chaos, the act of differentiating, identifies itself simultaneously with that of naming. For Thoreau it is a truly Adamic moment.

The generative language Thoreau describes is in fact preferable to an Adamic language, as organisms name themselves rather than being named from without, by a human act. Such a language cannot finally be lost but is always present and discernible to the careful observer, as Thoreau claims that "The earth is not a mere fragment of dead history, stratum upon stratum like the leaves of a book, to be studied by geologists and antiquaries chiefly, but living poetry like the leaves of a tree, which precede flowers and fruit,—not a fossil earth, but a living earth . . ." (309). The earth, then, is language itself: not dead, but living, both standard and copious. In this jubilant passage, Thoreau comes as near as possible to effacing himself as Author. The earth writes itself; Thoreau merely takes copy. His metaphor for language is finally not one of fluid sound, but of a living text, an object of sight, of leaves on the trees, not the sound of winds rustling through them. No longer arbitrary, language stabilizes meaning.

The forms in the melting bank flow only momentarily before they express a final form, a "hieroglyph" to be interpreted. Although the hieroglyph contains an element of the phonetic, its manifestation is nonetheless in the concrete, permanent form of writing.[3] In the end, the figure of the hieroglyph serves to check the unsettling dependence of the written word on the transitory moment of orality. In "Spring," flow becomes a final form, a seed developing to a predestined end, a greater stability. Language is finally determined and contained, much like the purely reflective water of Walden pond, without inlet or outlet.

In *The Maine Woods*, to the contrary, running waters and rapids threaten to carry the poet precipitously downstream. Likewise, symbol and significance threaten to escape, if indeed they can be captured at all. Before setting out into this new territory, Thoreau and his companion carefully copy a map of the region and trace, as he says, "what we afterwards ascertained to be a labyrinth of errors, carefully following the outlines of the imaginary lakes which that map contains" (15). What has been written or mapped does not signify the place and fails to match the topography. Similarly, on this first trip, Thoreau will discover unmapped psycho-linguistic ground on the summit of Ktaadn.

Thoreau's Ktaadn narrative serves as a counterpoint to the "Spring" chapter in *Walden* in that its inverted imagery undoes the tenuous synthesis between the spoken and the written word afforded in the earlier work and reintroduces the question of the mother tongue, the oral and time-bound element of language. Atop Ktaadn, Thoreau finds an entirely inorganic world of rock, silence, and clouds. Here he beholds not another melting bank, nor even a "crude mass of peat," but the unyielding bedrock of the world: "an undone extremity of the globe; as in lignite we see coal in the process of formation" (63). Although the vast chemistry of nature may promise the eventual fulfillment of Thoreau's poetic designs, its context belies such confidence in the present actual scene. Indeed, lignite forms as living matter and turns slowly into stone.[4] Here Thoreau's geologic metaphor reverses the process of "Spring"'s organic fecundity. The mountain is antithetical to the verdant valley and the human mind.

Atop Ktaadn, Thoreau discovers that Matter Earth is not Mater (Mother) Earth. There is no generative link between material and spiritual worlds as in the organic world of "Spring." Descending into the Burnt Lands, Thoreau can make neither psychic nor linguistic sense of his experience: "Perhaps I most fully realized that this was primeval, untamed, and forever untamable Nature, or whatever else men call it . . ." (69).[5] Even the term "Nature" fails to signify what he encounters on the summit of Ktaadn. The elemental has no name. "Here was no man's garden, but the unhandseled globe. It was not lawn, nor pasture, nor mead, nor woodland, nor lea, nor arable, nor waste land" (70). Thoreau's negative definition of the Burnt Lands implies their escape from comprehension, from the realm of language. It is no man's garden, it will not produce a term that has human value and meaning; in its extreme, it refuses even to be waste.

Ktaadn represents a critical absence, a central gap or silence where determining the significance of experience has become impossible. Where significance in "Spring" depends on the cessation of flow, its solidification into the articulate word, on Ktaadn flow becomes escape—the outward flow of the divine faculty between the ribs, the slippage between word and thing. Paradoxically, it is here on the solid earth that Thoreau confronts the unstable ground of language itself and becomes a stranger even to his own material body: "I fear not spirits, ghosts, of which I am one,—*that* my body might,—but I fear bodies, I tremble to meet them." The only reality is that available to the ghost within the machine. A skeptical abyss yawns before him: "Talk of mysteries!—Think of our life in nature,—daily to be shown matter, to come in contact with it,—rocks, trees, wind on our cheeks! the *solid* earth! the *actual* world! the *common sense*! *Contact*! *Contact*! *Who* are we? *where* are we?" (71). Dispossessed of the material world, the self engaged in and existing in language has become lost. As the melting bank passage eventually turned back to discover the organic derivation of its own words, so too this passage looks twice at its own language. In this case, however, the italicized words are remarkable for their failure to connect. What indeed is "solid" and "actual" in a world where mind has no "contact" with matter? Here sentences dissolve into exclamatory fragments resembling cries for help.

The central discontinuity of Ktaadn will remain the mystery, the pivot around which Thoreau attempts to find a way through the wilderness, even as geographically his later trips circumambulate the peak. In his later travels in Maine he follows both the path of the logger and the trail of the Indian in pursuit of the central symbols of the wilderness: the moose and the pine. These guides, however, do not follow the same North Star. These competing modes simultaneously raise questions about the ethics of the acts of logging and hunting, and of printing the language of the Other, be it the language of the Indian or of the wilderness.

The paradigmatic white man's way of being in the Maine wilderness is that of the logger. The logger's world is that of the written sign, of the marks Thoreau and his companions find on the logs stranded above the stream. The marks signify not only the absence of the owners, the makers of marks, but they also rely upon a second absence—that of the living tree. Once the logger inscribes his ownership, the pine is no longer a signifying tree but something quite different—a quantity of board feet. The logger's arbitrary mark signifies only the desire for possession. Thoreau holds that the poet may grasp and thus inscribe the deeper significance. We must, however, question Thoreau's affirmation of the poet's right to write "tree."

Thoreau's insistent assertion of the primacy of the poet masks his anxiety about the ethics of his own act of inscribing. Let us not forget that the paper the poet writes on also derives from trees (in fact, the poet comes to depend upon their demise), and that his act is also one of possession. Removed from their living source, the marks on the trees are bound for obliteration in the mill. Likewise, the production of the text risks falling prey to a purely material ethos, deprived of its spirituality and removed from experience by its association with commerce. In the "Allegash and East Branch" section, Thoreau comes near to realizing his own dilemma as a nature writer even as he condemns the misuse of the forest by the Anglo-American who "ignorantly erases mythological tablets in order to print his handbills and town-meeting warrants on them" (229). Affirming appreciation for the thing itself, Thoreau reveals his own implication in the process of sign making—that it is derivative and potentially inadequate and depends upon the removal of experience from the woods to the page. The Muses' Spring does not lie along the logger's path.

Turning from the logger's path to the Indian's trail, Thoreau seeks the "mother tongue" in the native tongue of the Penobscot Indians. His Indian education begins unpromisingly: in "Ktaadn" Thoreau's first reactions to Native Americans waver between a disappointed lament for the "corruption" of the Indian and a desire for contact with the romanticized, iconic Indian through whom it might be possible to reach the true meaning of nature. Although he and his companion bring their maps along on the second trip to the Maine woods, Thoreau hires Joe Aitteon to guide him both through the country and, more importantly, into it to provide inside knowledge the two-dimensional map cannot represent. The added dimension will appear as the motion of spoken language through time. Thoreau will not find translation as simple as he had at first anticipated.

This connection between language and nature remains elusive, even as Thoreau is unable to classify the "Indian" in any stereotypic role. The Indians we meet in *The Maine Woods* are not what we would expect from reading the "red face of man" passage in "Ktaadn." They are more Joe Aitteon and Joe Polis than "primitive men."[6] The inadequacy of Thoreau's preconceptions forces him into a challenging dialogue between his own writerly mode of being and the Indian's experiential mode. Although the Indians are more at home and better oriented in the woods than is Thoreau, their complexity as persons, their refusals to be read, insist upon a more dynamic model of language than that of the mystic "Indian Guide." When Polis states that he can teach Thoreau his language in a week, we recognize the gross hyperbole which reveals the impossibility of translating a life's experience through the shorthand of language. However, unexpected things can be learned by listening.

Through Aitteon and Polis, Thoreau seeks to supplement his vocabulary with Indian names for places, animals, and plants. In his philological pursuits Thoreau discovers a new, experiential type of language, a supplement which opens a larger and somewhat more treacherous linguistic field of play. The appendix to *The Maine Woods* reveals a multiplicity of meanings for many of its words. "Chesuncook," for example, means either "place where many streams empty in," "Big Lake," or "The Goose Place." Each use of the word reveals a different aspect, a new perspective on or experience of the object described. Meaning multiplies beyond the containable one-to-one relation.

Thoreau is thus caught between two worlds: by culture he is a logger, a maker of signs and possessor of meanings; by desire, he would be an Indian—or so he thinks. The Indian mode of being in the woods is that of the hunt—the search for the elusive moose. In the "Ktaadn" section Thoreau sees only the tracks of the moose; all the information he can offer the reader is hearsay. Noting places where the moose have left the distinct marks of their teeth, Thoreau remarks: "We expected nothing less than to meet a herd of them every moment . . ." (57). Unlike the marks made by loggers on the trees, however, those made by the moose imply the thing itself as an impending presence. Notebook in hand, Thoreau sets off in pursuit of the moose.[7]

The "Chesuncook" section involves Thoreau in the hunt, following the tracks to the conclusive experience of the moose, which, ironically, is its death. The Indian will take the moose's hide and flesh; Thoreau will turn the creature into prose. Thoreau collects and catalogues every available feature of the moose, visually dissecting it as he notes its color, its dimensions, "the delicacy and tenderness of the hooves," its awkward shape, the projecting upper lip, etc. He takes the ears for himself, perhaps as a totemic charm to help him discern the language of the wild. However, in this catalogue of the moose, the sum of its parts becomes less than the whole. His fascination ends in realization: "a tragical business it was,—to see that still warm and palpitating body pierced with a knife, to see the warm milk stream from the rent udder, and the ghastly naked red carcass appearing from within its seemly robe, which was made to hide it" (115–16). In the act of possession, moose becomes not-moose, a dead relic and empty sign.

While Thoreau deplores such wanton slaughter—"what a coarse and imperfect use Indians and hunters make of nature!" (120)—he cannot help but recognize his own complicity, even as he effaces it: "It is true, I came as near to being a hunter and miss it, as possible myself, and as it is, I think that I could spend a year in the woods, fishing and hunting just enough to sustain myself, with satisfaction" (119). Thoreau here proposes himself as the true Indian, untainted by the economic desires which lead Indians and hunters to kill moose only for their (salable) hides. In imagining himself a primitive man, he elides the violence necessary to life and to the making of signs. Subsequently, the "Allegash and East Branch" section withdraws its focus from the hunt, from the search for the "thing itself," to concentrate instead upon Polis and his knowledge of the woods—upon the means and context of sign making.

As his own maps and signposts prove increasingly inadequate, Thoreau notes that "[Aitteon] and Indians generally, with whom I have talked, are not able to describe dimensions or distances in our measures with any accuracy. He could tell, perhaps, at what time we should arrive, but not how far it was" (131). (The Indian's temporal notion of being in the woods reflects an experience of continual return to site, to camps, within the flow of time, which cannot be fixed as the written word can.) In contrast to the deceptive signpost to Chamberlain Lake, which leads Thoreau and his companion astray, Polis's "gazette" on the trunk of the axe-blazed fir tree serves as a mnemonic device representing continual return rather than possession. His sign, the bear paddling the canoe, "had been used by his family always" (199). Like the stories forever being retold around campfires, Polis's sign partakes of the ritualistic, repetitive mode of oral tradition, rather than the linear-progressive, consumptive mode of European writing.[8] It is a knowledge not easily possessed in a single week.

When asked how he guides himself in the woods, Polis responds with frustratingly inconclusive answers about looking at rocks or the slope of a hill. Thoreau speculates that a prelinguistic instinct guides the Indian, who "does not carry things in his head, nor remember the route exactly, like a white man, but relies on himself at the moment. Not having experienced the need of the other sort of knowledge, all labeled and arranged, he has not acquired it" (185). Being at home in the woods does not, in the end, require the ability to translate. Such knowledge is not universal, but local and circumstantial, bound to the immediacy of time and place. Context is everything; utterance, unnecessary.

In contrast to Thoreau, who confuses the moose's call with the sound of the axe, Polis can identify the moose by its call and can even speak its language. Thoreau revels in Polis's ability to speak with the musquash (muskrat): "I was greatly surprised,—thought I had at last got into the wilderness, and that he was a wild man indeed, to be talking to a musquash! I did not know which of the two was the strangest to me" (206). In his enthusiasm, Thoreau ignores the failure of the musquash imitation. This "speech" proves as deceptive as the "woodcutter's axe." The musquash ignores Polis, and for good reason, as Polis is calling him to his death. Polis's appropriation of the language of moose or musquash again separates sound from its "true" or "natural" origin. This capture of language not only removes sound from the source, changing its signification, but manifests a violent intent.

In *The Maine Woods*, word is inevitably separated from world by a fatal silence. In the "Chesuncook" section, Thoreau, wearied of the company of the loggers, joins the Indians at their campfire in what becomes for him a reenactment of the discovery of the New World. In feasting on the flesh of the moose in this "cannibalistic" scene, the Indians celebrate the lore of the moose's ways and of the hunt. The ritual death and consumption of the moose restores the Indian language to its pre-Columbian power, even as the Native Americans' respect for the animals they kill makes the moose live again in stories. As Thoreau listens in rapt self-forgetfulness to the language flowing around him, he momentarily abandons his attempt to decipher the Indian's gesture, to provide meaning for his readers.[9] Paradoxically, the original or native tongue is precisely that which cannot be translated: "There can be no more startling evidence of their being a distinct and comparatively aboriginal race than to hear this unaltered Indian language, which the white man cannot speak nor understand. We may suspect change and deterioration in almost every other particular, but the language which is so wholly unintelligible to us" (136). Although Thoreau's romantic yearning after an impossibly pristine language is still apparent, so too is his dawning recognition that it is death or absence which makes speech possible and which also maintains the deepest mystery by keeping the thing itself ever at remove from the possibility of possession. Thoreau has come full circle, back to the summit of Ktaadn and the inevitability of the death of the thing in the word, be it written or spoken.

How then can the writer represent nature while accounting for the violence speaking and writing necessarily performs? The answer may lie in what Thoreau begins to grasp in *The Maine Woods*, that however much is written, our contingent connection to the natural world takes the form of dialogue. Both for Thoreau and the reader he draws into this dialogue, the truth of the writing is realized only by a continual return to the woods, to the world, to check, reform, and revise what has been said, to continue the conversation now begun. The failure of language thus becomes not only its condition for existence but also part of its beauty and strength. Herein lies Thoreau's inability and final unwillingness to comprehend what lies beyond the first dark rank of trees, or indeed, the central mystery of Ktaadn itself, which he never climbs on a clearer day.

Notes

1. Garber follows Thoreau's reliance upon the written word and the concomitant failure of what is written in *The Maine Woods*—a move which enables his deconstructive reading of the limits of writing. The present essay takes a somewhat different turn in regarding Thoreau's encounter with the spoken word as equally important to his study of language in *The Maine Woods*.
2. For discussions of the instability of Thoreau's linguistic ground, see McIntosh and Garber.
3. Regarding significance of the hieroglyph in the American Renaissance, see Irwin.
4. Thoreau repeats this process of the transformation of life into stone in describing the small bird which "would flit away before me, unable to command its course, like a fragment of the gray rock blown off by the wind" (65). In the "stone sheep" passage, gray silent rocks replace pastoral herds (61). Thoreau's imagination is the only animating influence that allows them to stare back with hard gray eyes.
5. Tallmadge reads Ktaadn's "wilderness of words" in terms of a dual narrative structure with an initial, false climax atop Ktaadn, which conforms to the contemporary reader's expectations of the travel essay, and a second, true (anti)climax in the descent into the Burnt Lands, which disrupts and exposes the "melodrama, laced with clichés where language works to repress rather than express the narrator's experience" and replaces it with a naturalistic view (146).

6. For the ways in which the successive *Maine Woods* essays reveal Thoreau's changing atttitudes toward Native Americans, see Sayre, ch. 6.

7. Round sees Thoreau's explorations of Indian language as part of an attempt to escape the "overdetermined" discourse of natural history and notes that for Thoreau the Native American "straddles the materiality of common sense, and the irrationality of Chaos and Old Night" (321, 322).

8. Ong identifies the cyclic repetition of stories and traditions as the oral culture's mode of remembering. He also asserts that writing eliminates the need for remembering and allows a culture both to develop abstractions and to become self-conscious. The repetition of oral traditions does not entail a static system, as variation and difference are included in each new repetition, which, to the participants, remains "the same" in a way that writing-based cultures, seeking to identify terms, do not recognize.

9. See for comparision Greenblatt on early Spanish colonists' inability to securely attach meaning to their experiences of the New World. Murray notes that Indian gestures and signs were often seen as the least mediated form of communication and therefore closer to nature in their absence of syntax and abstraction. Thoreau, although assuming that the Indian gestures will help him translate, does not regard them as transparent means of accessing the discourse. His attempt is self-consciously comic, fanciful, and insufficient.

Works Cited

GARBER, FREDERICK. *Thoreau's Fable of Inscribing.* Princeton: Princeton UP, 1991.

GREENBLATT, STEPHEN. *Marvelous Possessions.* Chicago: U of Chicago P, 1991.

IRWIN, JOHN T. *American Hieroglyphics: The Symbol of Egyptian Hieroglyphics in the American Renaissance.* New Haven: Yale UP, 1980.

KRAITSIR, CHARLES. *The Significance of the Alphabet.* Boston: Peabody, 1846.

MCINTOSH, JAMES. *Thoreau as Romantic Naturalist.* Ithaca: Cornell UP, 1974.

MURRAY, DAVID O. *Forked Tongues: Speech, Writing and Respresentation in North American Indian Texts*. Bloomington: Indiana UP, 1991.
ONG, WALTER. *Orality and Literacy*. New York: Methuen, 1982.
ROUND, PHILLIP. "Gentleman Amateur or 'Fellow-Creature?': Thoreau's Maine Woods Flight from Contemporary Natural History." *Thoreau's World and Ours*. Ed. Edmund Schofield. Golden: North American P, 1993. 316–29.
SAYRE, ROBERT F. *Thoreau and the American Indians*. Princeton: Princeton UP, 1977.
TALLMADGE, JOHN. "'Ktaadn': Thoreau in the Wilderness of Words." *ESQ: A Journal of the American Renaissance* 31. 3. (1985): 137–48.
THOREAU, HENRY DAVID. *The Maine Woods*. Princeton: Princeton UP, 1972.
———. *Walden*. Ed. J Lyndon Shanley. 1854. Princeton: Princeton UP, 1971.

"*I only seek to put you in rapport*"
Message and Method in Walt Whitman's Specimen Days

DANIEL J. PHILIPPON

Perhaps "the most wayward, spontaneous, fragmentary book ever printed," Walt Whitman's *Specimen Days* (1882) consists of three parts: a short, autobiographical essay, written by Whitman for a friend in 1882; memoranda from Whitman's Civil War notebooks, written in and around Washington, D.C., from 1862 through 1865; and a combination of Whitman's nature notes and diary entries, written from 1876 through 1881, to which have been added miscellaneous essays and articles, including travel sketches, reminiscences, and discussions of prominent artists and writers.[1] Never the subject of much critical attention, the nature notes in *Specimen Days* have suffered doubly from the general disregard of the book as a whole and from the greater attention given to the biographical and Civil War portions of the text. In his introduction to the illustrated edition of *Specimen Days*, for example, Alfred Kazin describes the Civil War memoranda as comprising "the heart of the book" (xix) and proclaims the nature notes to be but "minor pastorals" (xxiv). "The picture of Whitman in the woods," Kazin says, ". . . does not have the excitement of *Walden* or 'Song Of Myself,' but it is altogether charming as an old man's last idyll" (xxiv).[2] Kazin and other critics underestimate the value of the nature notes in *Specimen Days* in part because they view the text as a passive object to be evaluated, rather than as an active attempt by Whitman to encourage engagement with the non-human world. Although the nature notes in *Specimen Days* are clearly an aesthetic creation, Whitman sees them more as a functional tool. Through them, he strives not to represent "nature" as an inanimate object for aesthetic consumption, but to "re-present" the nonhuman world as an animate landscape with which readers can and should have "rapport."

The writing of *Specimen Days* was itself a functional tool for Whitman, a form of therapy to help him recover from the devastating series of physical and emotional difficulties he faced in the decade following the Civil War. The first of these came in January 1873, when Whitman was partially disabled by a paralytic stroke on his left side, an injury that left him thoroughly depressed. Later that year he suffered the deaths of his sister-in-law Martha (in February) and his mother (in May), which further increased his depression. Unable to work, he took a leave of absence from his job in the Office of the Attorney General in Washington, joined his brother George in Camden, New Jersey, and moved into his mother's room, where—out of grief—he maintained the furnishings as they were before her death. A few months later, he lost his government post and, with it, his main source of income. Finally, in February 1875, at age fifty-six, he was crippled by a second stroke, this time down his right side (Kaplan 345–9). Whitman was, as he notes in *Specimen Days*, "quite unwell" during these years (118).

Whitman's outlook began to brighten in the spring of 1876 when he met eighteen-year-old Harry Stafford, who was working as an apprentice in the printing office of the *Camden New Republic*. Their friendship grew, and after a few weeks Harry took Whitman home to meet his parents, George and Susan Stafford, who were tenant farmers in Laurel Springs, New Jersey, twelve miles from Camden (Kaplan 359–60). The Staffords took a liking to the poet—"I think he is the best man I ever knew," Susan Stafford told Edward Carpenter—and they invited him to stay in their farmhouse at his convenience (Carpenter 11). Whitman accepted and thereafter lived with the Staffords as a paying guest for weeks at a time, remaining at the farm through the summer of 1876, and returning the following two summers (and sometimes in the winters) until George Stafford moved out of the house in March 1879 (Whitman, *Corr.* III:147). For therapy, Whitman would walk the few hundred yards down the farm lane to what he called "Timber Creek," a small stream that passed through the nearby woods, where he would sunbathe in the nude, take mud baths in an abandoned marl pit, and wrestle with saplings and tree limbs for exercise.³ "Every day, seclusion—every day at least two or three hours of freedom, bathing, no talk, no bonds, no dress, no books, no *manners*," Whitman recounts in *Specimen Days* (150); "it is to my life here that I, perhaps, owe partial recovery" (118).

In turning to nature for therapy, Whitman was helping to usher in the "back-to-nature" movement that swept the nation in the late-nineteenth century and affected nearly every aspect of life, including literature, art, architecture, education, recreation, and religion.⁴ As if writing a handbook for this new movement, Whitman opens the nature-notes section of *Specimen Days* by stating his belief in the fundamental position nature holds in human affairs: "After you have exhausted what there is in business, politics, coviviality, love, and so on—have found that none of these finally satisfy, or permanently wear—what remains? Nature remains; to bring out from their torpid recesses, the affinities of a man or woman with the open air, the trees, fields, the changes of seasons—the sun by day and the stars of heaven by night" (119–20). Nature, Whitman says, is "the naked source-life of us all . . . the breast of the great silent savage all-acceptive Mother" (122), "the only permanent reliance for sanity of book or human life" (120n). As a result, Whitman hopes his book may convey some of nature's graces to readers unable to experience the outdoors directly. "Who knows, (I have it in my fancy, my ambition,) but the pages now ensuing may carry ray of sun, or smell of grass or corn, or call of bird, or gleam of stars by night, or snow-flakes falling fresh and mystic, to denizen of heated city house, or tired workman or workwoman?—or may-be in sickroom or prison—to serve as a cooling breeze, or Nature's aroma, to some fever'd mouth or latent pulse" (120n).⁵

In attempting to capture the presence of nature in his pages, however, Whitman faces two challenges: he finds nature to be both "inimitable" (unable to be described) and wordless (unable to be interpreted). Throughout the nature notes, Whitman describes nature as "indescribable," an entity whose meaning can only be understood through participation with it.⁶ A rainbow he sees contains "an indescribable utterance of color and light" (144); a tulip tree near Timber Creek is "inimitable in hang of foliage and throwing-out of limb" (162); and butterflies have an "inimitable color" (178). As Stephen L. Tanner has observed, Whitman was a devoted stargazer, and many of his exclamations about nature's resistance to reproduction refer to the beauty of the sky. One "full-starr'd night" Whitman finds "every feature of the scene, indescribably soothing and tonic—

one of those hours that give hints to the soul, impossible to put in a statement" (147). Another night, he notes, "As if for the first time, indeed, creation noiselessly sank into and through me its placid and untellable lesson, beyond—O, so infinitely beyond!—anything from arts, books, sermons, or from science, old or new" (174). "Is there not something about the moon, some relation or reminder, which no poem or literature has yet caught?" he asks on a subsequent occasion (177). And, while watching the night sky reflected on the Delaware River, he exclaims, "Such transformations; such pictures and poems, inimitable" (186).[7]

Just as Whitman underscores the inability of words to capture the quality of his experience in nature, so too does he note the wordless quality of nature itself. Sitting alone near Timber Creek, he perceives "open, voiceless, mystic, far removed, yet palpable, eloquent Nature" (150). He describes the Milky Way as "some super-human symphony, some ode of universal vagueness, disdaining syllable and sound—a flashing glance of Deity, address'd to the soul. All silently—the indescribable night and stars—far off and silently" (174–5). He delights in "the sky and stars, that speak no word, nothing to the intellect, yet so eloquent, so communicative to the soul" (183). If linguistically silent, nature is nonetheless musically present. In the "deep musical drone" of bees, "humming their perpetual rich mellow boom," he wonders whether he hears "a hint . . . for a musical composition, of which it should be the back-ground? some bumble-bee symphony?" (124–5). He compares the movements of a Beethoven septette to nature: "Dainty abandon, sometimes as if Nature laughing on a hillside in the sunshine; serious and firm monotonies, as of winds; a horn sounding through the tangle of the forest, and the dying echoes; soothing floating of waves, but precisely rising in surges, angrily lashing, muttering, heavy, piercing peals of laughter, for interstices; now and then weird, as Nature herself is in certain moods" (233). And even when there is no wind, Whitman hears a "musical low murmur through the pines, quite pronounced, curious, like waterfalls, now still'd, now pouring again" (234–5).

Despite its wordlessness, nature still communicates with Whitman at Timber Creek, though it does so on a nonlinguistic level.[8] In the entry "The Oaks and I," for example, Whitman describes the "daily and simple exercise I am fond of—to pull on that young hickory sapling out there—to sway and yield to its tough-limber upright stem—haply to get into my old sinews some of its elastic fibre and clear sap. . . . I hold on boughs or slender trees caressingly there in sun and shade, wrestle with their innocent stalwartness—and *know* the virtue thereof passes from them into me. (Or may-be we interchange—may-be the trees are more aware of it all than I ever thought)" (152–3). The "rational" observer might dismiss such claims as "projections," but for Whitman this interchange between "The Oaks and I" is central to his recovery effort—and to his recovery narrative. Later in the same entry, he continues: "How it is I know not, but I often realize a presence here—in clear moods I am certain of it, and neither chemistry nor reasoning nor aesthetics will give the least explanation. All the past two summers it has been strengthening and nourishing my sick body and soul, as never before. Thanks, invisible physician, for thy silent delicious medicine, the day and night, thy waters and thy airs, the banks, the grass, the trees, and e'en the weeds!" (153).

Other entries reinforce the intersubjectivity of Whitman's experience in nature. When he hears the soft and pensive cooing of an owl during a break in his singing and recitations, Whitman fancies that the bird might be responding somewhat sarcastically to the style of his "vocalisms" (143). In another entry, Whitman recounts "a sort of dream-trance . . . in which I saw my favorite trees step out and promenade up, down and around, very curiously—with a whisper from one, leaning down as he pass'd me, *We do all this on the present occasion, exceptionally, just for you*" (162). Whitman makes a similar statement about the birds, noting that he has "a positive conviction that some of these birds sing, and others fly and flirt above here, for my especial benefit" (165).[9]

Given that nature is both inexpressible and wordless, Whitman must struggle to reconcile his understanding of nature's fundamental mystery with his desire to convey that mystery in language. In "To the Spring and Brook," an early entry, Whitman acknowledges the complexity of meaning-making, but also outlines his strategy for representing the natural world in his pages. Sitting by the side of a brook, he describes the sound of the water as "musical as soft clinking glasses" and records its "gurgling, gurgling ceaselessly—meaning, saying something, of course (if one could only translate it)" (121). Unable to offer an objective "translation" of the brook's "language," Whitman chooses instead to provide a subjective account of his own experiences at Timber Creek, hoping thereby to communicate some of nature's healthful influences to his readers. "Babble on, O brook, with that utterance of thine!" he declares. "I too will express what I have gather'd in my days and progress, native, subterranean, past—and now thee. Spin and wind thy way—I with thee, a little while, at any rate. As I haunt thee so often, season by season, thou knowest, reckest not me (yet why be so certain? who can tell?)—but I will learn from thee, and dwell on thee—receive, copy, print from thee" (121). In "The Lesson of a Tree," Whitman acknowledges the ability of nature to speak for itself, but he also suggests that texts such as his can help readers better understand the voice of nature:

> Science (or rather half-way science) scoffs at reminiscence of dryad and hamadryad, and of trees speaking. But, if they don't, they do as well as most speaking, writing, poetry, sermons—or rather they do a great deal better. I should say indeed that those old dryad-reminiscences are quite as true as any, and profounder than most reminiscences we get. ("Cut this out," as the quack mediciners say, and keep by you.) Go and sit in a grove or woods, with one or more of those voiceless companions, and read the foregoing, and think. (130)

Whitman's ultimate objective in writing *Specimen Days*, as he makes clear in this passage, is to render the text unnecessary, to have it serve as a vehicle to direct the reader's attention away from the written word and onto the "book of nature."

Whitman is certainly aware that texts have the power to shape human encounters with the nonhuman world. In the entry, "A Winter Day on the Seabeach," for instance, he describes his impressions from a day spent at the New Jersey shore, where poetry and music may have heightened the impact of the scene on his senses:

> That spread of waves and gray-white beach, salt, monotonous, senseless—such an entire absence of art, books, talk, elegance—so indescribably comforting, even this winter day—grim, yet so delicate-looking, so spiritual—striking emotional, impalpable depths, subtler than all the poems, paintings, music, I have ever read, seen, heard. (Yet let me be fair, perhaps it is because I have read those poems and heard that music.) (138)

At the same time, Whitman understands that texts have the ability to mislead the reader about "the *emotional* aspects and influences of Nature" (159). As he notes in one entry, "I, too, like the rest, feel these modern tendencies (from all the prevailing intellections, literature and poems,) to turn everything to pathos, ennui, morbidity, dissatisfaction, death. Yet how clear it is to me that those are not the born results, influences of Nature at all, but of one's own distorted, sick or silly soul" (159).

According to Whitman, therefore, the primary purpose of *Specimen Days* is not to create an admirable aesthetic object (although this may be a secondary purpose), but rather to put the reader in "rapport" with nature, to "receive, copy, [and] print" from nature in order eventually to direct the reader's attention to nature itself—"*first premises* many call it, but really the crowning result of all, laws, tallies and proofs" (294). Whitman's most detailed statement on the matter appears in the entry "After Trying a Certain Book," in which he offers a theory of interpretation for all "texts"—including music, poetry, and nature—and suggests that meaning is made through the interaction of sympathetic sensations.

> Common teachers or critics are always asking "What does it mean?" Symphony of fine musician, or sunset, or sea-waves rolling up the beach—what do they mean? Undoubtedly in the most subtle-elusive sense they mean something—as love does, and religion does, and the best poem;—but who shall fathom and define those meanings? (I do not intend this as a warrant for wildness and frantic escapades—but to justify the soul's frequent joy in what cannot be defined to the intellectual part, or to calculation.) . . .
>
> (*To a poetic student and friend.*)—I only seek to put you in rapport. Your own brain, heart, evolution, must not only understand the matter, but largely supply it. (292)

As Whitman states elsewhere in *Specimen Days*, nature is best understood through complete sensory participation, physical as well as mental: "Perhaps the inner, never lost rapport we hold with earth, light, air, trees, &c., is not to be realized through eyes and mind only, but through the whole corporeal body, which I will not have blinded or bandaged any more than the eyes" (152).

To bring about this "rapport" and thereby put the reader in sympathy with nature, Whitman adopts two techniques in *Specimen Days*: first, he employs a rhetoric of spontaneity, intimacy, and artlessness to stress the immediacy of animate nature; and, second, he structures his text as a series of discontinuous fragments to emphasize the ongoing, organic process of sensory perception.[10] This correlation between message and method may not always be obvious to the reader of *Specimen Days*, as Whitman no doubt was aware. In his introduction to the critical edition of *Specimen Days*, Floyd Stovall notes that "[t]he observations jotted down at Timber Creek with no other purpose than to express his sensations and reflections of the moment were later revised with an artist's care for the vivid effects, the complex rhythms, and the nice distinctions of thought and feeling that characterize his best prose writing" (vii). And as James E. Miller, Jr., comments in his review of Stovall's edition of the text, *Specimen Days* "strikes one at first as a grab-bag of miscellaneous jottings," but this first impression "gradually gives way to another impression: there is more method than at first appears" (90).[11]

Whitman attempts to convey the immediacy of his experience in three ways: he refers to the process of writing, he addresses the reader in the second person, and he describes his method of writing. Throughout the nature notes, Whitman frequently uses the phrase "as I write this" to give the impression of spontaneity. In the entry "Bumble-Bees," for example, he uses three versions of the phrase—"As I jot this paragraph," "As I write," and "As I write this"—as well as further variations on the phrase, such as "As I walk," "As I wend," and "as I return home" (123–5). In each of these locutions, the subordinating conjunction "as" helps to establish the temporal immediacy of Whitman's actions.

Whitman achieves this same effect another way by speaking to the reader directly with a term of endearment: "dear." In the "Interregnum Paragraph" separating the Civil War memoranda from the nature notes, Whitman expresses hope that "the notes of that outdoor life could only prove as glowing to you, reader dear, as the experience itself was to me" (118–9); "let me pick thee out singly, reader dear, and talk in perfect freedom, negligently, confidentially," he says in a subsequent note (122). Such turns of phrase not only foster an intimacy between Whitman and the reader, but they also link the activity of reading to Whitman's own activity, making the reader a co-participant in the immediacy of the author's experience.

Finally, Whitman attempts to make his text appear artless by explaining his method of composition to the reader. In the first of his nature notes, for instance, he suggests that his apparent lack of editing should keep the reader focused on nature and not art, because the best art comes the closest to nature's formlessness. "Literature flies so high and is so hotly spiced that our notes may seem hardly more than breaths of common air, or draughts of water to drink," Whitman claims. "But that is part of our lesson" (120). A few introductory notes later he declares, "But to my jottings, taking them as they come, from the heap, without particular selection.... Each was carelessly pencilled in the open air, at the time and place" (122). And in the penultimate entry of the book, Whitman returns to this theme, explaining that "there is a humiliating lesson one learns, in serene hours, of a fine day or night. Nature seems to look on all fixed-up poetry and art as something almost impertinent.... [A]fraid of dropping what smack of outdoors or sun or starlight might cling to the lines, I dared not try to meddle with or smooth them" (293).

The corollary to Whitman's emphasis on the present moment is his use of discontinuous fragments to convey the unfinished quality of sensory perception. This paratactical method addresses two limitations of perception as Whitman sees them: the inability to perceive all of nature at once, and the inability to perceive nature directly. Like his memoranda of the war, Whitman's nature notes resemble brief journalistic bulletins from the front. No one can perceive all of nature at once, just as no one can experience the whole war at once, so Whitman chooses instead to provide specimen moments of his experience in nature, much as he sees the rapid flight of a flock of geese as providing "a hint of the whole spread of Nature" (234). Moreover, Whitman recognizes that our lives in nature are themselves fragmentary and unfinished, a series of "specimen days" that often go unheralded. "[H]ow few of life's days and hours (and they not by relative value or proportion, but by chance) are ever noted," he observes (1). "Whitman does not view truth as the sum of all experience," as Linck C. Johnson puts it, "but rather as the knowledge both he and his reader gain from apprehending the intense meaning of these particular moments, and seeing the pattern formed by these moments" (8). Thus, one of Whitman's goals in rendering his text as a series of moments is to remind the reader that "no two places, hardly any two hours, [are] anywhere, exactly alike" (235). Each is unique, and each must be experienced in its full expressiveness.

Another of Whitman's goals is to offer a way around the "indirections and directions" involved in the perception of nature (138). If the indescribability and silence of nature pose a problem for Whitman in his attempt to put the reader in rapport with the natural world, one way he seeks to overcome this problem is through the accumulation and repetition of moments. As he notes in the entry "A

Discovery of Old Age," "Perhaps the best is always cumulative. . . . In my own experience, (persons, poems, places, characters,) I discover the best hardly ever at first, (no absolute rule about it, however,) sometimes suddenly bursting forth, or stealthily opening to me, perhaps after years of unwitting familiarity, unappreciation, usage" (277–8). Many of the phrases Whitman repeats throughout the nature notes suggest this repetitive method of his text. For instance, he often uses the phrase "long and long," as in "here I sit long and long" (124), "I stopp'd long and long" (190), or "I stood long and long" (267–8). The titles of three of his entries also reflect this quality of repetition: "Crows and Crows," "Birds and Birds and Birds," and "Mulleins and Mulleins." Some of the names Whitman rejected in favor of "Specimen Days" further reveal the repetitive nature of his thinking: "Week In and Week Out," "Flood Tide and Ebb," "Far and Near at 63," "Fore and Aft . . . . . Vestibules," and "Again and Again" (248n). Faced with the problem of a silent, indescribable subject, Whitman remains undaunted, "again and again" addressing the issue in the hope that by repeated attention, the richness of his life in nature might somehow be revealed.

A late work often overlooked, *Specimen Days* is in fact Whitman's most thorough consideration of the relationship between humans and their nonhuman environment. "The most profound theme that can occupy the mind of man," he writes in *Specimen Days*, ". . . is doubtless involved in the query: What is the fusing explanation and tie—what the relation between the (radical, democratic) Me, the human identity of understanding, emotions, spirit, &c., on the one side, of and with the (conservative) Not Me, the whole of the material objective universe and laws, with what is behind them in time and space, on the other side?" (258). Borrowing his formulation of the question from Emerson and Carlyle, Whitman first suggests that the statements of German idealist philosophers, and particularly Hegel, might offer "the last best word that has been said upon it, up to date" (259). But then Whitman reconsiders and declares that philosophy finally cannot equal poetry in its ability to convey the animate, passionate quality of the human-nature relationship:

> While the contributions which German Kant and Fichte and Schelling and Hegel have bequeath'd to humanity—and which English Darwin has also in his field—are indispensable to the erudition of America's future, I should say that in all of them, and

the best of them, when compared with the lightning flashes and flights of the old prophets and *exaltés*, the spiritual poets and poetry of all lands, (as in the Hebrew Bible,) there seems to be, nay certainly is, something lacking—something cold, a failure to satisfy the deepest emotions of the soul—a want of living glow, fondness, warmth, which the old *exaltés* and poets supply, and which the keenest modern philosophers so far do not. (260–1)

Clearly, Whitman aligns himself with those "old *exaltés* and poets" who are able to supply the "living glow, fondness, [and] warmth" of human experience in nature. Proof of this characterization appears in an August 1879 letter Whitman wrote to John Burroughs, in which he enclosed a two-paragraph comment about himself (written in the third-person) for inclusion in an essay Burroughs was writing about "Nature and the Poets." Part of the comment reads: "Through all that fluid, weird Nature, 'so far and yet so near,' he finds human relations, human responses. In entire consistence with botany, geology, science, or what-not, he endues his very seas and woods with passion, more than the old hamadryads or tritons. His fields, his rocks, his trees, are not dead material, but living companions" (Barrus, *Whitman* 111). No doubt in agreement with Whitman's own description of himself, Burroughs included the passage almost verbatim in his essay.[12]

Notes

1. Walt Whitman, *Prose Works 1892*, vol. 1: *Specimen Days*, ed. Floyd Stovall (New York: New York UP, 1963) 1. Subsequent references to this volume will be made parenthetically. In "Withdrawal and Resumption," William Aarnes says the text has four parts (402); George B. Hutchinson says it has five (4). The Civil War notebooks, originally published as *Memoranda During the War* (1875), were included in *Specimen Days* with slight revisions.
2. In a review of Kazin's edition of *Specimen Days*, Leo Marx also describes the Civil War memoranda as comprising "the heart of the book" (6). In the foreword to the 1961 New American Library edition of *Specimen Days*, Richard Chase likewise states that "[a]s a nature writer Whitman can scarcely be said to rank with the best; he is no Thoreau" (xiv). Not every critic has been so dismissive, though few have studied the text in depth. In a late October 1882 letter, John

Burroughs told Whitman he liked the nature notes (Barrus, *Life* I: 248); Gay Wilson Allen says the notes are "as intimate and personal as one can find in romantic prose" (154); and Walter Teller proclaims the nature notes to be "some of the best prose he [Whitman] ever wrote" (2). Justin Kaplan and Bettina L. Knapp provide more recent, slightly longer treatments. The most thorough commentary to date on the nature notes in *Specimen Days* appears in the dissertation of William Aarnes, parts of which were published in four journal articles. See, in particular, "'Cut This Out,'" in which Aarnes argues that Whitman seeks "to put his reader in touch with Nature as an embodiment of the spiritual realm" (26). In contrast, I stress Whitman's emphasis on the animate physicality of nature.

3. Herbert Gilchrist's 1878 sketch of Whitman sunbathing in the nude, except for a a hat and shoes, appears as the frontispiece to vol. 6 of the *Correspondence*.

4. See Peter J. Schmitt, *Back to Nature*. Roderick Nash also claims that "Whitman was a precursor of the American celebration of savagery" at the turn of the century (151), and Harold Aspiz documents the ways in which Whitman's regimen paralleled the therapeutic air cures and camp cures prescribed by his physician, S. Weir Mitchell, and other doctors in the 1870s and 1880s.

5. In this and subsequent paragraphs, I quote extensively from *Specimen Days* to demonstrate the two terms of my title: the message of the text and the method by which it is conveyed. In the final portion of this essay I discuss Whitman's rhetorical techniques in detail.

6. Whitman makes a number of similar statements in the Civil War section of *Specimen Days*. He claims the sky is "expressively silent" (67); he describes the call of the cattle drivers in Washington as "indescribable" (68); and, perhaps most famously, he says that "the real war will never get in the books" (116). Similar sentiments appear in the entries "Soldiers and Talks," "A Night Battle, over a Week Since," "Unnamed Remains the Bravest Soldier," and in the note to "A Happy Hour's Command." Whitman also finds President Lincoln to be "inimitable"; see the entries "Abraham Lincoln" and "No Good Portrait of Lincoln."

7. Whitman clarifies the manner in which nature is "inimitable" in the entry "Summer Sights and Indolences," in which he notes that the skies are indescribable "in quality, not details or forms" (127). In other words, Whitman can describe objects and events, but not their overall impact.

8. I disagree with Glenn N. Cummings, therefore, who argues that "[t]he relationship is not one of two unified spirits—the poet's with Nature's—but the distanced relationship an audience has with a theatrical performance. It is not a union, but rather a spectacular, entertaining disunion" (182).

9. In the Civil War section, Whitman similarly asks, "The Weather.—Does It Sympathize with These Times?": "Whether the rains, the heat and cold, and what underlies them all, are affected with what affects man in masses, and follow his play of passionate action. . . ?" (94).

10. The issue of Whitman's method in *Specimen Days* must inevitably raise the question of genre. Joseph Eugene Mullin turns to the *silva*, a verse collection of variety and seeming spontaneity, to explain the "casual ordering" and "fragmentary nature" of *Specimen Days* (156–9). And although Edward E. Chielens classifies *Specimen Days* as part of the genre of the familiar essay, he admits that Whitman's use of repetition "places the work on the outer boundaries of the genre" (376). I am inclined to side with Chielens in seeing the text as part of the long tradition of the personal essay, which, beginning with Montaigne, seeks the kind of intimacy with the reader Whitman intends by his use of the term "rapport." The nature essay, of course, is itself a species of this genre.

11. William White's "Author at Work: Whitman's *Specimen Days*" offers a glimpse of Whitman's editing of one entry; Stovall's edition provides variant readings of the text and includes an appendix of omitted passages.

12. The essay appears in Burroughs's *Pepacton*, 85–126; the passage appears on 119–20.

Works Cited

AARNES, WILLIAM. "Scraps: A Study of Walt Whitman's *Specimen Days*." Ph.D. diss. Johns Hopkins U, 1979.

———. "'Cut This Out': Whitman Liberating the Reader in *Specimen Days*." *Walt Whitman Review* 27 (1981): 25–32.

———. "Withdrawal and Resumption: Whitman and Society in the Last Two Parts of *Specimen Days*." *Studies in the American Renaissance* (1982): 401–32.

ALLEN, GAY WILSON. *The New Walt Whitman Handbook*. New York: New York UP, 1975.

ASPIZ, HAROLD. "*Specimen Days*: The Therapeutics of Sun-Bathing." *Walt Whitman Quarterly Review* 1.3 (1983): 48–50.

BARRUS, CLARA. *The Life and Letters of John Burroughs*. 2 vols. New York: Houghton, 1925.

———. *Whitman and Burroughs: Comrades*. Boston: Houghton, 1931.

BURROUGHS, JOHN. *Pepacton*. Boston: Houghton, 1881.

CARPENTER, EDWARD. *Days with Walt Whitman*. New York: Macmillan, 1906.

CHASE, RICHARD. "Foreword." *Specimen Days*. By Walt Whitman. New York: New American Library, 1961. ix–xvi.

CHIELENS, EDWARD E. "Whitman's *Specimen Days* and the Familiar Essay Genre." *Genre* 8 (1975): 366–78.

CUMMINGS, GLENN N. "Whitman's *Specimen Days* and the Theatricality of 'Semirenewal.'" *American Transcendental Quarterly* 6 (1992): 177–87.

HUTCHINSON, GEORGE B. "Life Review and the Common World in Whitman's *Specimen Days*." *South Atlantic Review* 52 (1987): 3–23.

JOHNSON, LINCK C. "The Design of Walt Whitman's *Specimen Days*." *Walt Whitman Review* 21 (1975): 3–14.

KAPLAN, JUSTIN. *Walt Whitman: A Life*. New York: Simon and Schuster, 1980.

KNAPP, BETTINA L. *Walt Whitman*. New York: Continuum, 1993.

KAZIN, ALFRED. Introduction. *Specimen Days*. By Walt Whitman. Boston: D. R. Godine, 1971. xix–xxiv.

MARX, LEO. "Specimen Days." *New York Times Book Review* 21 November 1971: 6+.

MILLER, JAMES A., JR. "Notes for an Autobiography." Rev. of *Specimen Days*, by Walt Whitman, ed. Floyd Stovall. *Walt Whitman Review* 9 (1963): 90–2.

MULLIN, JOSEPH EUGENE. "The Whitman of *Specimen Days*." *The Iowa Review* 24 (1994): 148–61.

NASH, RODERICK. *Wilderness and the American Mind*. 3d ed. New Haven: Yale UP, 1982.

SCHMITT, PETER J. *Back to Nature: The Arcadian Myth in Urban America*. 1969. Baltimore: Johns Hopkins UP, 1990.

TANNER, STEPHEN L. "Star-Gazing in Whitman's *Specimen Days*." *Walt Whitman Review* 19 (1973): 158–61.

TELLER, WALTER. "Speaking of Books: Whitman at Timber Creek." *New York Times Book Review* 10 April 1966: 2+.

WHITE, WILLIAM. "Author at Work: Whitman's *Specimen Days*." *Manuscripts* 18.3 (1966): 26–8.

WHITMAN, WALT. *The Correspondence*. Ed. Edwin Haviland Miller. 6 vols. New York: New York UP, 1961–77.

———. *Memoranda During the War*. Camden, NJ: Author's publication, 1875–76.

———. *Prose Works 1892*. Vol. 1: *Specimen Days*. Ed. Floyd Stovall. New York: New York UP, 1963.

PART IV
READINGS OF TWENTIETH-CENTURY
ENVIRONMENTAL LITERATURE

Beyond the Excursion
Initiatory Themes in Annie Dillard and Terry Tempest Williams
JOHN TALLMADGE

Annie Dillard's *Pilgrim at Tinker Creek* and Terry Tempest Williams's *Refuge* are two of the most powerful works to appear in the current renaissance of American nature writing. Each dramatically extends the possibilities of the genre, not only by virtue of style and substance, but also by virtue of theme and structure. In their use of the excursion format, which has informed nature writing for more than two centuries, Dillard and Williams open exciting but quite different paths for transforming our relations to nature.

The local excursion, or "ramble," as Thomas J. Lyon terms it in his taxonomy of nature writing (4–5), begins in Gilbert White's *Natural History of Selbourne* as a simple neighborhood walk during which the curious naturalist merely records observations. A religious and psychological dimension enters with William Wordsworth, for whom nature provided lessons in the conduct of life and the motions of the mind. Henry Thoreau combined these two approaches to create a deliberate practice in which walks followed by reflective journalizing provided scientific awareness, moral and psychological understanding, and prophetic social critique. This complex Thoreauvian excursion has been most influential in our tradition in which nature has often been viewed as a source of religious revelation or a standard of moral value.

Assumptions about nature, protagonists, and their reciprocal relations are implicit in this Thoreauvian model. Nature is conceived as both knowable and meaningful in human terms. Beings in nature stand forth as signs, whether of spiritual truths (Ralph Waldo Emerson, Edward Abbey), natural laws (Joseph Wood Krutch, Loren Eiseley), or the history of creation (John Muir, Aldo Leopold). The world thus revealed accords with human nature, either by corresponding to structures of thought and feeling or by reflecting our chosen values. The protagonist engages this world as an observer or learner with an attitude that is contemplative as opposed to aggressive, manipulative, or consumptive. This attitude reflects a benign, essentially harmonious relation between self and world.

Much of *Pilgrim*'s revolutionary impact comes from the way Dillard plays with these conventions. The book is a series of excursions arranged on the seasonal cycle of an idealized year, a pattern common in American nature writing; its title signals its local grounding and religious emphasis, both of which reinforce our generic expectations. Yet the first episode is not an excursion but an *in*cursion where the boundaries between nature and self are rudely violated. The old fighting tom cuts a shocking, paradoxical figure that blends domesticity with violence, eroticism, and mystery. He leaps in without warning, assaulting the passive narrator and leaving her body marked with vivid, inscrutable signs. Yet the cat is a pet, not a wild animal, or so we would like to think. It's as if a hidden and possibly dangerous aspect of things suddenly reveals itself. As for the narrator, her reaction is equally strange. One would expect her to jump out of bed with a shriek, but she only "half-awakens" as the cat massages her bare chest with his bloody paws. It's almost as if she likes it—a not altogether comforting thought—and a few lines later we learn that this episode is no fluke: she sleeps with the window open, and the cat's visits take on an air of ritual ("our rites," 2). This groggy, expectant narrator has a faintly decadent air about her, suggesting the somnambulant heroines of late romantic novels who begin to crave visitations from the vampire. By using scriptural analogues to interpret the cat's visit ("the keys to the kingdom or the mark of Cain," 2), Dillard suggests that his paradoxical and disturbing qualities may betoken aspects of God's own nature. There is something creepy yet fascinating in a situation that combines eros and domesticity with violence, danger, and vulnerability while throwing over the whole a coloring of religion.

The rest of the book plays dazzling variations on the theme of this keynote passage: that nature reveals a God who is powerful, fascinating, violent, and inscrutable. Dillard goes out into nature seeking encounters with power and mystery. "I am no scientist," she says. "I explore the neighborhood" (12), but she is interested mainly in the "unmapped dim reaches and unholy fastnesses" of speculation to which these local experiences lead her (11). She construes the walk as a religious practice that engages beings in nature as tokens of God's character and intentions. It's a type of mystic journey or vision quest carried out in the backyard.

Dillard's experiences seem to fall into two broad categories. In some cases, she's stunned by the beauty of nature, especially by light, color, energy, and motion. She watches light shows on the trees and cliffs, pats the puppy in the parking lot, sees the "tree with the lights in it, . . . each cell buzzing with flame" (35). These encounters reveal God as an extravagant, joyous creator, a magician who dazzles and baffles his delighted audience. They fill Dillard with rapture; she feels like a bell that rings after being struck. In evoking these experiences, Dillard relies on metaphors couched in active verbs that suggest both the transformative power of revelation and the awe felt by the observer:

> Shadows lope along the mountain's rumpled flanks; they elongate like root tips, like lobes of spilling water, faster and faster. A warm purple pigment pools in each ruck and tuck of the rock; it deepens and spreads, boring crevasses, canyons. As the purple vaults and slides, it tricks out the unleafed forest and rumpled rock in gilt, in shape-shifting patches of glow. These gold lights veer and retract, shatter and glide in a series of dazzling splashes, shrinking, leaking, exploding. The ridge's bosses and hummocks sprout bulging from its side; the whole mountain looms miles closer . . . .(79)

Here each metaphor is vividly precise, and the rhythm is perfectly modulated, but there is too much information to interpret consciously. So we respond intuitively, with a sense of brightness and excitement, just as the narrator responded to the scene before her. The intricate language and syntax correspond to the complex order and presentation of nature itself. By locating most of the metaphors in verbs, the strongest words syntactically and grammatically, Dillard heightens the sense of drama and energy. The scene is hard to visualize but easy to feel, and it is also easy to feel good about.

In other cases, however, Dillard is shocked by nature's perversity, grotesqueness, or cruelty. The giant water bug sucking life from the frog, the lethal intimacies of parasites, the ichneumon eggs that hatch inside and devour their mother, the praying mantis who munches down her lover, all these challenge her faith in a humane and loving God. Although the most vivid examples come secondhand, through her reading in naturalists like Edwin Way Teale and Henri Fabre, she finds disturbing corroborations during her own suburban walks. These fascinate and repulse her, violating both her morals and the decorum of the excursion mode. *Pilgrim* caused a sensation in part because of Dillard's insistence on giving equal time to aspects of nature that seem perverse and cruel, things many of her predecessors had eschewed. For her nature is as much a horror show as a light show.

The emotion of horror arises from the sudden awareness of danger in a place of security. It's an emotional lurch from comfort to fear and loathing. Reading a Poe short story, for instance, we might be drawn in by the eloquent, charming narrator only to learn in the very last line that he's really a psychopathic axe murderer. In Stephen King novels, familiar objects that we take for granted, like cars or cornfields, become bloodthirsty predators. Similarly, in *Pilgrim*, small, unobtrusive creatures that we habitually overlook are revealed as signs of a mysterious and potentially monstrous divinity. The effect is rather like finding a rattlesnake under your pillow. Dillard's keynote example is the giant water bug sucking the frog. The passage begins on a note of comic familiarity, with Dillard setting off like a mischievous kid to scare frogs, expecting them to jump with a "Yike!" like spooked Kermits. But when one frog doesn't react this way, she realizes that something is wrong:

> . . . he slowly crumpled and began to sag. The spirit vanished from his eyes as if snuffed. His skin emptied and drooped; his very skull seemed to collapse and settle like a kicked tent. He was shrinking before my eyes like a deflating football. I watched the taut, glistening skin on his shoulders ruck, and rumple, and fall. Soon, part of his skin, formless as a pricked balloon, lay in floating folds like bright scum on top of the water. . . . An oval shadow hung in the water behind the drained frog. (6)

Here the chain of similes draws out the moment of fear and fascination, building tension as the narrator tries desperately to interpret. It isn't until the frog is completely sucked out that she realizes the "oval shadow" is a giant water bug. Her similes are all drawn from domestic life: snuffed candles, tents, footballs, balloons. They suggest childhood and play, all in accord with the opening images. But into this world comes the giant water bug, deadly serious with its hooked legs and lethal enzymes. It "hugs its victim tight" as it injects the poison, like a perversion of maternal or spousal embrace. Dillard clinches the horror by noting matter-of-factly that "this event is quite common in warm fresh water" (6). In other words, it could happen in your neck of the woods and you never noticed, did you? She makes the reader feel as vulnerable as the frog.

This passage is perfectly symmetrical. From a comfortable domestic image of nature we pass through a moment of horror (mediated by similes) to an image of nature as powerful and cruel. The effect on both narrator and reader is shock: "I couldn't catch my breath," says Dillard as she gets up from her knees (7). She might have been praying or taking a beating: the experience batters her mind and heart as they cling to the idea of a humane and merciful God. Dealing with this kind of assault is the principal aim of her book.

Despite their differences in tone, these two kinds of experiences have one thing in common: they provide intense sensations that take the protagonist into a state of extreme consciousness. In this respect, exaltation and horror are two sides of a coin, a fact well known to poets like Byron and Rimbaud. The result in either case is visionary rapture, and this is what the pilgrim goes into nature to seek. Dillard wants to get as close to the power that moves the universe as she can without dying or going mad. In her practice, the excursion is transformed from an observer's ramble in benign, Selbourne-like environs to a quest for powerful sensations in a strange and perilous landscape. This sort of pilgrimage entails risk and suffering. Dillard's imagery suggests painful, initiatory transformations: "some enormous power brushes me with its clean wing, and I resound like a beaten bell" (13); "I stood with difficulty, bashed by the unexpectedness of this beauty, and my spread lungs roared" (41); ". . . you'll come back, for you will come back, transformed in a way you may not have bargained for—dribbling and crazed" (277). Powerful assertions like these accumulate throughout the book, yet oddly enough there is little evidence of change in the character or behavior of the protagonist. She seems as constant as the northern star, an impression reinforced by the book's cyclic, seasonal structure and coda of final paragraphs that echo the opening encounters with the tom cat and the giant water bug. While the language suggests the suffering and transformation of initiation, the action itself suggests repetition and stasis, as if the initiation were not completed or the experiences were being sought for stimulation rather than growth. One begins to suspect Dillard of indulging in a kind of "romantic agony," seeking pain and rapture in order to feed her imagination. The difference between this sort of suffering and the growth pains of initiation becomes clearer when we turn to Terry Tempest Williams.

Like *Pilgrim at Tinker Creek*, Williams's *Refuge* presents itself as a story of suffering and transformation. It is episodic and dramatic, consisting in large measure of local excursions. Unlike *Pilgrim*, however, its narrative is linear, covering a seven-year period during which the Great Salt Lake rises to inundate the Bear River Migratory Bird Refuge and Williams loses both her mother and grandmother to cancer. In *Pilgrim* excursions alternate with meditative or explanatory passages developed out of the narrator's reading; the format reflects the fact that Dillard composed from a pile of index cards in the library (*Writing* 30). In *Refuge*, the excursions alternate with scenes in civilization: home, the sick room, the hospital. These worlds provide experiences of searing intensity; in fact, the most painful occur in town with family members, particularly her dying mother, rather than out in nature, although the Great Basin offers a much harsher environment than Dillard's Blue Ridge.

For Williams the excursion is initially a means of escape, first from the claustrophobia of urban life and the gender expectations of Mormon patriarchy, and later from the grinding anguish and helplessness of nursing cancer victims. She does not go to nature seeking intense sensations—her life is dishing up more than enough of that—but rather to find peace, hope, and healing for herself. Her excursions are seldom undertaken alone. Unlike Dillard, she goes out with friends, family members, or colleagues. So her journeys not only teach her about the land in its scientific and spiritual dimensions, but also deepen her relationships with people; they are attempts to connect, to compensate for the separation and loss brought on by flooding and cancer. Whereas in Dillard the excursion unfolds as a binary transaction between nature and the solitary observer, in Williams the transaction is a complex, three-way encounter involving at least two persons interacting with the land and one another. Williams's narrative, therefore, is always embedded in a context of social relations. Nature's lessons apply to pressing personal issues that are also politically charged. In this respect *Refuge* takes the excursion beyond a self-centered romanticism toward the socially transforming process Thoreau envisioned in "Walking."

Williams's initial journeys to the Bear River Migratory Bird Refuge and Great Salt Lake affirm both her difference from the norms of Mormon womanhood and her closeness to her female mentors, her mother Diane, and her paternal grandmother, Mimi. From Diane she takes a love of solitude and a forceful temperament; from Mimi, a love of birds and a sense of the hidden, spiritual dimension to things. Both women maintain an inner spirit of independence and integrity while observing the decorum of Mormon culture. When the book opens, Williams is basing her sense of identity upon the twin foundations of family and place, and her excursions affirm that sense of self. But as cancer and flooding commence to erode and finally destroy these foundations, her excursions become more challenging. She moves from one kind of suffering to another, from the agony of the sick room to the ferocity of salt, sun, and wind, the desolation of flooded marshes, and the arrogant intrusion of technology in the form of giant pumps designed to suck floodwaters into the remote West Desert. The later excursions bring pain but also wisdom conveyed by signs—the appearance of certain birds, the discovery of ancient artifacts, an accident that leaves her scarred. Eventually, they also bring moments of healing, so that the suffering becomes redemptive rather than just an experience of nature's power.

In *Pilgrim*, the protagonist goes into nature seeking visions, and what she sees causes shock and anguish. But she remains an observer, like Emerson's transparent eyeball, curiously detached from what she sees. Her sufferings all take place in her mind, in the solitude of her thought and reading. In *Refuge*, however, the protagonist engages bodily with nature: she's chafed by the wind, burnt by the sun, stung by the brine of Great Salt Lake. The story is punctuated with images of extraordinary physical intimacy that evoke both pain and healing. After her friend Tamra Crocker Pulsifer dies of cancer, Williams finds a dead swan by the lake and is moved to prepare it for burial:

> The small dark eyes had sunk behind the yellow lores. It was a whistling swan. I looked for two black stones, found them, and placed them over the eyes like coins. They held. And, using my own saliva as my mother and grandmother had done to wash my face, I washed the swan's black bill and feet until they shone like patent leather. (121)

This sacramental act carried out in a harsh, unforgiving environment is at once pathetic and radiant with hope. Somehow ministering to the swan salves the inner wound that Williams suffers from not being able to heal her family and friends. By embracing the harshness of the desert she finds opportunities for connection and insight: "In the forsaken corners of Great Salt Lake there is no illusion of being safe. . . . Only the land's mercy and a calm mind can save my soul. And it is here I find grace" (148). After her mother's death she retreats to a cave in the salt desert for healing (237); the earth seems to be cradling her, and we think of her own attempts to heal and comfort her mother when she lies down in the sick bed and starts breathing with her as she massages her back, as one might soothe a young child (157). In this womb-like place, sanctified by a dripping spring and ancient pictographs, Williams grieves unreservedly: "My keening is for my family, fractured and displaced" (238). And shortly thereafter, on a trip to southern Utah, Williams falls down a canyon wall and splits her forehead open, receiving from the land a physical wound corresponding to the wound in her psyche:

> It ran from my widow's peak straight down my forehead across the bridge of my nose down my cheek to the edge of my jaw. I saw the boney plate of my skull. Bedrock.
>
> I have been marked by the desert. The scar meanders down the center of my forehead like a red, clay river. A natural feature on a map. I see the land and myself in context. (243–44)

Whereas what Dillard sees makes her suffer, what Williams sees comes to her through suffering. Pain and grief cleanse her vision, destroying false hopes and illusions. It's an involuntary ascetic process that intensifies as the cancer and flooding progress. As she gives up her cherished illusions (symbolized at one point by the mirage), she begins to receive signs in the form of dreams, intuitions, and animal manifestations. For example, after the second operation her mother confesses that she had never been allowed to grieve for her childhood dog, and just at that moment a black lab (the identical breed) trots up out of nowhere (163). Similarly, her grandmother has owl premonitions a week before her death, even though she has never seen owls in the city, and just after she passes away Williams goes out into the backyard and finds two owls who stare at her before flying off (272). The owl is considered a spiritual emissary and omen of death in some Native American cultures, and its appearance at this most significant moment confirms the "clandestine vision of things" that Williams shares with her grandmother (273). She is now certain that life and nature have a spiritual dimension and that relationships continue beyond death, "something," she confesses, "I did not anticipate" (275).

This ascetic process whereby illusions are stripped to make way for insight is given expressive form in the style and format of *Refuge*. Williams's spare, lean prose draws attention to the events themselves rather than to the narrator's art. The story is told as a succession of vivid, epiphanic moments that are strung rather than woven together. The sense of drama and unmediated revelation is thus intensified while the sense of the narrator's self-absorption, so dangerous to intimate autobiography, is reduced. In *Pilgrim* the narrator's art stands forth on every page, not only conspicuous but dazzling. Most of Dillard's finest insights are mediated if not entirely created by the play of language. But in *Refuge* the narrator's art recedes into the background, creating the sense that it is not Williams herself but the events she has witnessed that address us. *Pilgrim* is full of memorable, even famous lines; like Emerson, Dillard is highly quotable. But one does not easily quote *Refuge*; one remembers instead the events, and the story into which they cohere. *Refuge* is not quotable but tellable; its voice is not that of a preacher but a storyteller.

The story of *Refuge*, I would suggest, is that of a true initiation. Whereas Dillard's sufferings are largely mental and momentary, those of Williams are visceral and protracted. She faces the stress of major threats to body and soul—cancer, the loss of a parent and grandparent, the violation of her home landscape—and she survives all this without going mad. In the process she becomes stronger. And she changes from an innocent, rebellious child to a mature woman, the matriarch of her family and a voice for change in her culture. In *Pilgrim*, Dillard's anguish and rapture both result from her extraordinary attentiveness. One might say she projects a "heroics of sensitivity." But in *Refuge* the suffering is inflicted by chance, fate, and the fallen world's capacity for evil. Williams endures it, determined to draw out its meaning. Hers is a "heroics of witness."

Williams's own transformation from daughter to matriarch is reflected in a series of changes in the excursion mode itself. Initially, her journeys resemble the classic ramble: she's a bird watcher cruising the refuge and descrying her quarry through binoculars. With the advent of disease and flood, the excursion becomes an escape, but nature too is revealed as wounded and impermanent. As her anguish intensifies, Williams's journeys lead to increasing physical intimacy with the land. She goes seeking comfort but receives signs instead, and that begins a process of acceptance and healing. Eventually, she begins to make ceremonial gestures of acknowledgment, as when she arranges the dead swan in the shape of a cross, places stones on its eyelids, and cleans its beak with her own saliva (121). By the final episodes the excursion has become a ritual, as when she takes her grandmother to visit the Sun Tunnels or paddles out on the lake with her husband to scatter marigold petals in memory of her mother (267, 280). Ritual is one of the most ancient human techniques for the management of change; it affirms the meanings that endure beyond the depredations of history. In this particular case, ritual both sustains Williams's relation to her ancestors and the land through death and devastation, and articulates the new mode of being into which her experiences have propelled her.

Williams's transformation is truly initiatory while Dillard's only appears so. Williams, as protagonist, moves through suffering into a new mode of being, whereas Dillard's relation to self and world never changes despite the intensity of her sensations and visions. For Dillard the initiation is not completed but suspended, as in the case of Coleridge's Ancient Mariner, who returns with a "strange power of speech" but cannot escape his ordeal. Dillard's final image suggests this sort of possessed narrator: "And like Billy Bray I go my way, and my left foot says 'Glory,' and my right foot says 'Amen': in and out of Shadow Creek, upstream and down, exultant, in a daze, dancing, to the twin silver trumpets of praise" (279). It's a more joyous image than that of the Mariner—more like the dancer of "Kubla Khan" perhaps—but revealing in its consistency with Dillard's earlier celebrations of pentecostal ecstasy. The protagonist ends her pilgrimage as a solitary dancer; there's no one in the picture but her and the Spirit, masquerading, of course, as nature. And the dance seems to have no purpose other than praise, no outcome other than its own beauty. It feels, on reflection, like a performance undertaken for its own sake, dazzling but self-sufficient in its grace.

Contrast *Refuge*, which ends with an act of civil disobedience as Williams and a group of women enter the Nevada Test Site to protest the atomic weapons whose lethal artifice has made both her family and place "unnatural." We see now that Williams's initiatory transformation is meant to counteract the demonic transmutation of nature by nuclear science and industrial culture. To their physical power she opposes the spiritual power of a matriarch with a prophetic mission to bear witness and so redeem her culture. That her final excursion should be an act of political protest realizes the potential for nature writing sensed by Thoreau and articulated most recently by Barry Lopez, who imagines that the genre might one day provide "the foundation for a reorganization of American political thought" (qtd. in Halpern 297). From excursion to initiation to witness, such is the progress of nature writing in our time. *Pilgrim at Tinker Creek* opens the door, and *Refuge* shows us the way.

Works Cited

DILLARD, ANNIE. *Pilgrim at Tinker Creek*. New York: Bantam, 1975.
———. *The Writing Life*. New York: HarperCollins, 1989.
HALPERN, DANIEL, ED. *On Nature*. San Francisco: North Point, 1987.
LYON, THOMAS J. *This Incomperable Lande: A Book of American Nature Writing*. Boston: Houghton Mifflin, 1989.
WILLIAMS, TERRY TEMPEST. *Refuge: An Unnatural History of Family and Place*. New York: Pantheon, 1991.

Aimé Césaire's A Tempest *and* Peter Greenaway's Prospero's Books *as Ecological Rereadings and Rewritings of* Shakespeare's The Tempest[1]

PAULA WILLOQUET-MARICONDI

The world is all around me, not in front of me.
MERLEAU-PONTY

Aimé Césaire's *A Tempest* (1969) and Peter Greenaway's *Prospero's Books* (1991) are theatrical and cinematic "rereadings" and "rewritings" of Shakespeare's *The Tempest* that express our contemporary concerns with the power of science and technology to alter and destroy the ecosystemic relations on which hinges the survival of all beings. These works examine the effects of dis*place*ment on both Being and Environment by reprioritizing the interconnectedness and interdependency of Self and Place. Both artists manipulate the Shakespearean text in order to suggest a new "ethic of being" founded on the "recovery of an effective identifying relationship between self and place" (Ashcroft 8).

These two works invite us to contemplate the correspondences between the seventeenth century and our own time. As Greenaway explains in an interview, Shakespeare's *The Tempest*, being a play about beginnings and endings, "is perhaps very relevant to the end of the century, the end of the millennium" (Rodgers 12). Césaire's and Greenaway's renditions of the play look back to the origins of the current ecological crisis and forward to the urgent need for reversal of ecological destruction. For these artists, the seventeenth and twentieth centuries represent crucial moments in the history of humanity's relation to nature. These periods are marked, respectively, by science's displacement of animism and alchemy and by the rise of the modern instrumentalist ethos on the one hand, and, on the other, by the questioning of that ethos and by the need to recover or rediscover a nonexploitative and participatory relationship with the natural world. Césaire's and Greenaway's works echo the appeals made three decades ago by Lynn White Jr., and more recently by Max Oelschlaeger, to rediscover, or at best to reinvent the notion of the Sacred—a Sacred that is, however, not necessarily rooted in any contemporary religious doctrine.

In Césaire's version, Prospero's survival on the island depends on his willingness and ability to exchange relations of domination for adaptation and integration. If Prospero is to be seen as a representative of the West, then Césaire's decision to leave Prospero on the island may be interpreted as a challenge posed to the Western audience to redefine its own sense of relation to place. In Greenaway's version, Prospero is trapped in his own creation, the story we know as *The Tempest*, which he authors in the course of the film. Prospero's liberation, as well as that of the island and its inhabitants, comes only with the surrendering of techniques—Prospero's art and magic. When Prospero's source of power is nullified, so is the product of that magic. As Prospero, aided by Ariel, hurls his twenty-four books into the sea and breaks his magic stick in half, he frees not only himself but the entire island and its inhabitants from his despotic grip. The final breakdown and collapse of his creation—an illusion-become-reality—is accompanied by a return of the natural sounds of the island. *Prospero's Books* thus constitutes an exploration and a critique of modernity's intention to "author" the world as a human artifact, and its misguided faith in the quest for absolute understanding and control of human and nonhuman nature. The stylistic and thematic modifications Césaire and Greenaway bring to the Shakespearean text are profoundly political acts that attempt to initiate a "decolonization of the mind" by presenting a *new* non-canonical reading of an *old* canonical text (Césaire, *Discourse* 78).

In adapting the play for a black theater, Césaire casts Caliban as a black slave and Ariel as a pacifist mulatto slave; he also enlarges the relationship between Caliban and Prospero by allowing Caliban to fight back. Moreover, Césaire introduces a new character, the Yoruba trickster god-devil Eshu, and gives a voice and a presence to both the human and the nonhuman native inhabitants of the island: the animals and insects, the island's spirits, the Wind, and Shango, the force of thunder and lightning.

The modifications Césaire brings to the play must be understood in light of Césaire's elaboration, in 1935, of the concept of *négritude*. The term *négritude* was designed to invoke a sense of belonging to an ethnic community, of being part of a specifically black African history which had been denied to West Indians, like Césaire, by colonialism and slavery. For Césaire, *négritude* was a sociocultural ideology. In 1969, at the time *A Tempest* was written, Césaire's position was that black culture existed, but that it was a historical fact and not a biological one. In 1984, Césaire declared that *négritude* was a means to reach and uncover an authentic self and a sense of heritage and place which the Martinican author lacked, his relationship to the African land and history having been mediated by colonialism (see Arnold). His work, particularly his poetry, has thus been seen as a quest for this heritage and for a sense of self in relation to place.

Césaire's version of *négritude* is striking in its emphasis on the relations between self and place, an emphasis which is fundamental to his poetry as well as to *A Tempest*. For instance, when positing the black man's identity, Césaire proclaims in *Notebook of a Return to the Native Land* (1939), "my *négritude* is neither a tower nor a cathedral / it plunges into the red flesh of the sun / it plunges into the burning flesh of the sky." Césaire's denunciation of colonization and his celebration of an identity rooted in a sense of intimacy, respect, and partnership with place may be seen in the following verses, also from *Notebook*: "And we are on our feet now, my land and I / with windswept hair, my small hand / now in its enormous fist" (trans. by and qtd. in Arnold). What these verses suggest, and what *A Tempest* attempts to stage, is the fact that the decolonization project necessitates a rethinking of our relations with the land. A rebirth of the cultures oppressed by the West entails a rebirth of the land, and land must be conceived not as property or mappable space, but as a form of being in its own right.

This conception of place as being, and as sacred, is part of the African animistic world-view linking mysticism and a profound sense of collective belonging described by Maurice Delafosse, an influential scientific popularizer of Africa in France in the 1920s. The divinity in animistic religions manifests itself through place; thus, place is sacred. As Lynn White Jr. reminds us, Western religions' denial of animism and the sacredness of the natural world makes possible the heedless exploitation and domination of nature (1206). In *A Tempest*, this concept of divinity is present in the characters of the Wind, of Shango, Eshu, and the island's spirits—characters that symbolize a resurrection of African spirituality.

The animistic world-view Césaire found in African mysticism is today being expressed by proponents of deep ecology, an environmental movement that promotes the belief in a community of being that comprises living and nonliving entities, of which living humans as well as the next generation and the dead are members (see Devall and Sessions, *Deep Ecology*). As Bill Devall succinctly puts it, deep ecology is "premised on a gestalt of person-in-nature" (303). For animists and deep ecologists alike, "knowledge" of the world is sensually and intellectually identificatory. We are both a part of and a participant in the world, and self-realization must be recognized as a communal process, not an individualistic event. As Paul Shepard reminds us, "environment" is as much within as without, and the separation between self and other, between interior awareness and exterior stimulus, is merely the convention according to which our civilization has defined its relationship to the world.

This gestalt of being and environment is dramatized by both Césaire and Greenaway in their treatment of Ariel. When Ariel is first introduced, we learn that he has been clamoring for his freedom ever since Prospero "freed" him from the pine in which Sycorax had imprisoned him. In Césaire's text, Ariel grieves for his lost unity with the tree, saying that "after all, I might have turned into a real tree" (13). From an animistic or deep ecological perspective, this passage dramatizes the colonizer's role in alienating self from place. Prospero's "freeing" of Ariel is a form of dis*place*ment, analogous to the sundering of spirit from nature brought about with the destruction by Christianizing authorities of the sacred groves of European paganism.

The tree is an important symbol for Césaire because it represents the stability and attachment to place which are being lost through colonization, territorial expansion, and environmental degradation. In an essay entitled "Poetry and Cognition," Césaire celebrates the tree and its superiority to humans who have lost their sense of fraternity and integration with Place. For Césaire, the tree is superior because it represents fixity, attachment, and perseverance in what is essential. The crucial role that Césaire has always attributed to the poet in helping us regain a sense of rootedness lies in the fact that for him the poet, "like the tree, like the animal, surrendered himself to primal life; he said yes, he consented to that immense life which surpassed him. He rooted himself in the earth; he stretched out his arms; he played with the sun; he became a tree; he blossomed; he sang" ("Poetry and Cognition" 119). For Césaire, poetry offers us hope of a return to origins, for it calls upon our unconscious in which is preserved our original unity with nature.

In uprooting Ariel, Prospero becomes the opposite of the poet; for Césaire, he is the man of science who wants to know the world so as to have control over it. In "Poetry and Cognition," Césaire opposes scientific knowledge to poetic knowledge, explaining that, while scientific knowledge abstracts, classifies, categorizes, and explains, it does not reveal "the essence of things" (112). Thus, Césaire writes:

> Poetic cognition is born in the great silence of scientific knowledge. Through reflection, observation, experimentation, man bewildered by the data confronting him, finally dominates them. Henceforth he knows how to guide himself through the forest of phenomena. He knows how to use the world. But that does not make him king of the world. Image of the world. Yes. Science can offer him an image of the world, but briefly and superficially. . . . Scientific knowledge enumerates, measures, classifies and kills. To acquire it man has sacrificed everything: his desires, fears, feelings and psychological complexes. (112)

Césaire makes Prospero the target of his critique of imperialism and scientism and assigns the privileged role of poet to Caliban, the black slave. Whereas Prospero sees the island as empty and mute, a blank page on which to inscribe his monologic master narrative, Caliban celebrates its aliveness, his integration within it, and his dialectical relationship with it. Caliban's desire to "get back my island and regain my freedom" reminds us that his ability to be free is inextricably bound to the freeing of the land (69). Regaining a sense of community with Place is thus an important step in the decolonizing enterprise. The ending of the play reveals an aged, weary, and weak Prospero who, determined to stay on the island to "protect civilization," risks being defeated by the very nature he worked so hard to contain, and which now reclaims its territory. The play closes with the island and its native elements reasserting their presence and reclaiming their rightful Place. "It's as though the jungle was laying siege to the cave" (*A Tempest* 75).

While Césaire substantially modernizes the Shakespearean text to reflect not only his personal vision but the concerns of the historical moment in which he is writing, Greenaway remains faithful to the language of the original text, which the characters speak in the course of the film. Greenaway introduces important modifications, however, in relation to the twenty-four books Prospero takes into exile and in relation to Prospero himself. While Shakespeare merely mentions these books as the source of Prospero's power, Greenaway makes them the structuring device of the film. These books purport to comprise a collection of all knowledge about the world, and the film dramatizes not only the fabricated nature of this knowledge but the despotic potential of accumulating knowledge and equating knowledge with power. Through the use of sophisticated high-definition technology, Greenaway visually animates these books, bringing their content back to life and challenging the notion that the world can be reduced to accumulable units of knowledge which can be contained in texts, drawings, charts, and maps.

Like Césaire, Greenaway questions the feasibility and desirability of the rationalist and scientific pursuit of knowledge. The film dramatizes the fact that the acquisition and use of knowledge are never neutral, objective, or impartial, but always motivated, self-serving, and reductive. Knowledge is a discursive practice; to "know" the world is not to uncover its secrets, but to construct a theory, or narrative, about the world which impoverishes it and reduces its ambiguities. As Umberto Eco once noted, being "is always a hypothesis posed by language," and the world constructed from conjectures is only "logically" more satisfying than the organic world (252). To know the world this way is to mistake representations of the world for its organic beingness. What makes Shakespeare's play relevant to our times, explains Greenaway, is that "it's a play about knowledge and the uses of knowledge. We are all so knowledgeable now and there's so much knowledge available, that in some senses, we, too, have become magicians" (qtd. in Turman 106).

Among Prospero's twenty-four books there is an unfinished collection of thirty-six plays by a W. S. in which several pages have been left blank for the inclusion of one last play, *The Tempest*, to be written by Prospero. As the film begins, so does Prospero's authoring of the story of revenge which we are about to see. The characters Prospero "creates" as he scripts his tale of revenge are like words in a discourse; they lend their bodies to the creation of meaning in Prospero's signifying chain. They are quite literally prisoners of Prospero's device, shackled to his narrative. Prospero's revenge is thus an instrument, a technique of his will-to-author; with every technique comes "the potential to remake the world in harmony with the human imagination, so that every desire might be fulfilled" (Oelschlaeger 90). What is unique about Greenaway's cinematic rendition of the Shakespearean drama is that, in the text that Prospero fathers, the characters are denied their own voices and are ventriloquized by Prospero himself. Prospero, played by Sir John Gielgud, not only authors the characters' lines but speaks for them throughout most of the film, thus embodying the tyranny of authorship by retaining total control over the characters' performances.

Greenaway's Prospero epitomizes the modern visionary subject, creator and master manipulator of a fiction-become-reality; he is the Cartesian *cogito* who, like Descartes, and Plato long before him, negates the sensual—the organic, dynamic, and unpredictable reality of the island and its inhabitants—only to "construct the visible according to a model-in-thought" (Merleau-Ponty 169). Prospero's "scriptural enterprise" (de Certeau 135) is, to begin with, one of conquest and colonization, and then one of geographical and historical rewriting: he first appropriates the island and its inhabitants in order to rewrite them so as to create a replica of his renaissance kingdom and mentality; he then rewrites the past, as well, through a story of revenge designed to redress the wrongs which he feels have been done to him. As Stephen Tyler reminds us, "the true historical significance of writing is that it has increased our capacity to create totalistic illusions with which to have power over things or over others as if they were things. The whole ideology of representational significance is an ideology of power" (131). Unable, or rather unwilling, to integrate himself within the given natural space of the island, Prospero transforms this space into his own territory, rendering it speakable, that is, representable in his colonizer's language.

Prospero's power is his ability to abstract; his most powerful tool of control is language—specifically, written language, which introduces a sense of order and linearity to the world that replaces the cyclical rhythms of nature and the fluidity of the spoken word. This linearity is reflected in the narrative Prospero constructs when he begins to author the story which we know as *The Tempest*.

Prospero's use of language as a technique of abstraction for the creation of totalistic illusions grants him a concrete power and authority over people, things, and places. Under Prospero's linguistic control, the environment ceases to be a given and begins to be a product of his creation. Michel de Certeau calls writing the "fundamental initiatory *practice*" which posits the existence of a distinct and distant subject and a blank space or "page," in which "the ambiguities of the world have been exorcised." Prospero deliberately mistakes the organic island on which he lands for "the island of the page," a blank and empty space waiting to be scripted (135). Prospero, of course, also imposes his own language on the inhabitants of the island, thus confirming a fact already understood in 1452 by Antonio de Nebrija, Queen Isabella's grammarian: that language is the "companion" of empire (Greenblatt 563). The establishment and imposition of a national language, argues de Certeau, "implies a distancing of the living body (both traditional and individual) and thus also of everything which remains, among the people, linked to the earth, to the place, to orality or to non-verbal tasks" (138–39).

Through the use of language, Prospero *orders* the world in two senses: he *summons* a world into being and *imposes* a specific structure on it. He assigns to himself the position of authority, thus establishing a clear hierarchy between himself and his "creations"—the characters he conjures to effect his revenge, and the spirits and inhabitants of the island. This ordering process, which establishes and maintains Prospero's position of authority within a hierarchy, amounts to what Kenneth Burke calls a "pyramidal magic," a concept which for Burke is inevitable in social relations and a product of alienation and reification (qtd. Wolfe 82). The concept is visually rendered in the film by the presence of several pyramids and obelisks, the most dominant pyramid resembling the Pyramid of Cestius in Rome. These geometrical and rigid structures made of gray stone, marble, or terra-cotta brick are, we are told, "like pyramids that have been enthusiastically built on the hearsay evidence of travellers . . . that have been constructed by an antiquarian like Prospero who obtained his knowledge from books, not first-hand observation" (Greenaway 98). The pyramids stand between the golden, geometrical maze-like cornfield in the foreground and the dark forest beyond the horizon, separating Prospero's tamed and "civilized" domain from "wild nature."

Greenaway calls Prospero a "book-making-machine" whose knowledge is either derived from the books he brings with him to the island or manufactured by him and assembled into books. His is a knowledge-at-a-distance rather than the kind of knowledge, or understanding, one gains from unmediated interaction with the world. Prospero adopts an instrumentally rational attitude toward the island which nullifies it and strips objects and beings of any immanent purpose. Reduced to a blank space, and then repopulated with creatures from Prospero's books and imagination, the island becomes a reflection of Prospero's mind, a concretization of his own vision—his thoughts and hallucinations. Prospero's imaginings are always reflected in mirrors within the film, thus dramatizing the fact that they are *reflections* of his consciousness. The island can now be seen as a symbol of Prospero's own ego and of his nonparticipating consciousness.

Prospero's separation from the world is his method for mastery; it is a strategy, in de Certeau's sense of the term (35–38). For de Certeau, a strategy, unlike a tactic, requires that the mastering subject, in this case Prospero, situate himself in a place from which he will exert control—his cell. Prospero thus defines his position as being *outside* that which he intends to dominate. This strategy affords him the illusion of immunity and independence from the variability of circumstances and from chance; prediction and control are his primary goals. This method, argues de Certeau, is managerial and scientific; it embodies the Cartesian attitude that calls for a place of power, a totalizing discourse, and a compartmentalization of space. Prospero's cell is a locus, not an environment. From this position Prospero can develop a totalizing narrative which constructs the world as readable and representable.

Like Césaire's Prospero, Greenaway's Prospero is an alchemist become scientist whose tools and methods, like those of modern science, afford him a supernatural power over the world. Alchemy was a "science" of participation which acknowledged the dialectical nature of reality and promoted a respectful "intervention" in the world. Until the fifteenth century, any form of intervention or meddling in nature was accompanied by religious ceremonies designed to acknowledge the sacredness of nature. Starting in the sixteenth century, a paradigm shift occurs: the displacement of alchemy by science, culminating in the 1700s in the thorough discrediting of alchemy. The Scientific Revolution breaks with this participatory consciousness and achieves the previously unrealizable alchemist's dream: the transmutation of the physical and biological world into a machine that could be studied, understood, and controlled.

Max Oelschlaeger argues that "the modern mind has lost any sense of human dependence on an enveloping and therefore transcending *source of life*" (338). Prospero's egocentric—rather than ecocentric—relation to the island evokes the modernist self who lives in isolation from the world. Morris Berman addresses this point when he says that modern persons see themselves in atomistic terms, as islands, whereas their ancient or medieval predecessors embodied a more holistic view of self and environment (77). For the latter, the self was like an embryo, immersed in and continuous with its environment; self and world were seen as parts of the same body. To the extent that Prospero "creates" the island according to his Renaissance model, he *is* an island in Berman's sense, separate and unitary. Prospero is a disembodied intellect, an image strikingly rendered in the final moments of the film when Prospero's "talking head" appears on the screen against a black background. Prospero is nothing more than a mind in a void, growing smaller and smaller as the close-up image retreats from the viewer. The total absence of a body reminds us that Prospero's involvement with the world has, all along, been purely intellectual rather than physically and sensually participatory.

This participating consciousness described by Oelschlaeger and Berman was identified by Plato as "pathological," a diagnosis which justified his banning of poetry from the republic. "The poetic, or Homeric mentality," Berman explains, "in which the individual is immersed in a sea of contradictory experiences and learns about the world through emotional identification with it (original participation), is precisely what Socrates and Plato intended to destroy" (71). As Césaire has shown, poetry's ability to induce emotional identification and fusion with the experiential world, and to evoke multiple and contradictory experiences, prevented the self from becoming unitary and unified.

The organicity and materiality of the body and its interdependence with place, which Prospero negates and the poet celebrates, is most vividly foregrounded through the figure of Caliban, played by dancer and choreographer Michael Clark. The central role given to Caliban's body in the film serves to suggest an alternative means of playfully experiencing the world and constitutes an antidote to Prospero's hegemonic rationalism. Caliban's close association with both water and earth connotes an identificatory relation to the world, rather than an intellectual examination and domination of it. Caliban's movements draw our attention to the centrality of the body in the acquisition of tacit knowledge. He remains closely associated with water, the film's symbol of the source of life from which everything originates and to which everything returns.

Caliban is first seen emerging from the murky brownish waters of the pit to which he has been "exiled" by Prospero for attempting to violate Miranda. As Prospero speaks Caliban's lines, Caliban's body begins slowly and ambiguously to reveal itself. He is not quite human, not quite animal. Although not a fish, he is said to be suited to water and to snort like a hippopotamus. Particular attention is drawn to his colorful genitals and his small, curled horns. As Prospero speaks *for* him, Caliban's body crawls around the rock to which he is bound in dance-like contortions. Throughout the film, all of Caliban's movements are executed in this same dance-like fashion, continuously emphasizing his nakedness and dramatizing the physicality of his being.

The space of the pit where Caliban lives offers a startling contrast to that of Prospero's cell, the realm of knowledge, rationality, and control. Much like Plato's cave, Caliban's pit is portrayed as a primitive, debased, and dangerous world of shadows and deceptions. The light coming through the circular opening in the pit gives the characters and objects inside a ghostly and primal aura, while also suggesting the presence of an "enlightened" world outside the pit. In the pit, as in Plato's cave, it is the sensible world that takes on spectral qualities. Caliban, his mother Sycorax, and the creatures that dwell in the pit are, according to Prospero, associated with the dark powers of nature and black magic while Prospero's "magic" is said to be benign.

In the film, Caliban and Sycorax are shown to be closely bonded to the natural elements of the island. Greenaway retains the Shakespearean description of Sycorax as a witch "so strong / That she could control the moon, make flows and ebbs, / And deal in her command, without her power" (Greenaway 159). Caliban's fusion with nature is rendered by his constant evocation of and interaction with the natural elements of the island, his rightful "kingdom" until Prospero's arrival. As Caliban reminds Prospero, and us:

> And then I loved thee,
> And showed thee all the qualities o'the isle,
> The fresh springs, brine-pits, barren place, and fertile.
> Cursed be I that did so! All the charms
> Of Sycorax, toads, beetles, bats, light on you!

> For I am all the subjects that you have,
> Which first was mine own kind! And here you sty me
> In this hard rock, whiles you do keep from me
> The rest o'th'island. (Greenaway 94)

Caliban's symbiotic relationship with the environment is stressed once again at the end of the film when Caliban emerges from a body of water, an enveloping medium that dramatizes the "undividedness of the sensing and the sensed" (Merleau-Ponty 163). However, Greenaway assigns to this closing image an ambiguous meaning, for Caliban emerges from the water in order to rescue the two volumes Prospero has thrown in the sea. This rescue is necessary to the logic of the film's narrative since it is, presumably, thanks to Caliban's recovery of the books that we come to know the story of *The Tempest* as told by Greenaway. But this rescue invites at least two possible readings: are we to take Caliban to be a new "messiah" who, by bringing *The Tempest* to our attention, invites us to rethink our relationship to the natural world? Or are we to take the recovery of the books as the first step in a renewed cycle of logocentrism and domination of nature? Since we are not shown what Caliban ultimately does with these books, and since the film we have just watched effectively forces us to revisit *The Tempest* in a new light informed by ecological considerations, it seems that Greenaway does not intend for the answer regarding the fate of these books to lie in the film, but in us. We are thus assigned the task of resolving the ambiguous ending in a way commensurate with the film's ecological tone.

The passage from fusion with nature to separation brought about by Prospero's rationalistic and discursive model of interaction with the world, and the pain associated with such separation, are further dramatized in the sequence in which Ariel is "freed" from the pine tree. Whereas Ariel's fusion with the tree is taken to be a form of imprisonment by Prospero's nonparticipating consciousness, the image lends itself, as we have seen in Césaire's version, to a more holistic interpretation. While Prospero's relation to the world is atomistic, the image of Ariel in the tree suggests an embryonic interconnectedness with the environment of which he is a part and on which he depends. Ariel's fusion with the tree can be taken as evidence that person and environment are not only profoundly connected but, as Morris Berman argues, "ultimately identical" (77). This image captures the reversibility of experience between self and world whereby to experience the world is to be experienced by it. This intertwining represents a mode of being alternative to our metaphysical detachment from the sensible and living world.

Ariel's emergence from the tree, like the emergence from the sea of the shipwreck victims, is a form of birth into Prospero's symbolic world. This birth is marked by pain, and it leaves scars. Prospero deliberately misidentifies the origin of the pain, however, locating it in Ariel's "imprisonment" in the tree rather than in his "liberation" from it. The painful severing of Ariel from the tree, graphically rendered by the film, is described in the screenplay as clearly a sundering process:

> Ariel in the pine tree is set free. With a loud, painful scream—pulling and stretching and the sound of ripping. . . . Ariel emerges from the bark, moss and lichen . . . his body bloody. He stands in the light—moaning with pain . . . the blood running from around his neck, his armpits and groin . . . as we watch . . . the running blood disappears and his body—white in the light—is faintly marked with tree-rings . . . these too slowly vanish. (Greenaway 89)

This passage not only underscores Ariel's birth into Prospero's symbolic world but, through the repetition of the word "light," also stresses the "enlightened" nature of this separation from the natural world. As Prospero sees it, Ariel's liberation from the tree is a coming-into-the-light of reason; it is an event analogous to leaving the Cave. If the tree rings are emblematic of Ariel's communion with the tree—and thus with the whole of nature—the gradual disappearance of this gentle but crucial evidence attests to the fragility not only of this bond, but of our memory of it. Prospero's demonization of Sycorax, his banishing of Caliban, and his redefinition of Ariel's fusion with the tree in negative terms are the fictions he constructs to validate his rewriting of the past and of the island, and to assure his absolute power and control over all.

In their rewriting of the Shakespearean text, Césaire and Greenaway show that with the advent of science, an animistic worldview is replaced by the Cartesian mechanistic paradigm, still operative today, which dictates that our interaction with the world be one of confrontation rather than mutual permeation. What was once animated and living becomes lifeless and mechanistic; the rooting out of animistic belief is seen as a process of maturation—in other words, as progressive. The material world is stripped of its soul and reduced to matter, data, phenomena which can be confronted, studied, and "known." The Cartesian separation of mind and body is extended to the world at large. The now "disembodied intellect," incarnate in the figure of Prospero, no longer exists in a symbiotic relationship with the world. For the disembodied mind, nature is nothing more than a separate measurable and quantifiable theoretical object of inquiry.

Césaire and Greenaway dramatize the fact that Prospero's defeat lies in his violation of what Stephen Tyler calls "the first law of culture, which says that 'the more man controls anything, the more uncontrollable both become'" (123). As the representative of the imperialistic and scientific project, Césaire's Prospero is ultimately defeated by his inability to accompany the island's natural movements, while Greenaway's Prospero willingly drowns his tools of power and relinquishes control over the island. The different resolutions each artist brings to the drama may well reflect the historical contexts in which they were writing—the 1960s and the 1990s, respectively. While Césaire's resolution may seem optimistic in its suggestion that nature will eventually take over and restore itself in spite of humanity's efforts to destroy it, it is also pessimistic—though perhaps realistic—in its inability to grant humans the will to change. Greenaway's ending is more consonant with the imperatives we face today to put down our weapons, break our "magic sticks," and stop the mass destruction of environments and species, including the human species.

These different conclusions notwithstanding, the modifications Césaire and Greenaway bring to the characters of Prospero, Ariel, and Caliban are crucial in establishing these works' animistic and ecological stance and in denouncing the illusion of control and permanence encouraged by the totalizing scientific enterprise. These works' emphasis on integration and wholeness and their call for an interdependence of Place and Being suggest a healthy vision for the present and the future of humans' relations to one another and to the world at large. Césaire's and Greenaway's rereadings of *The Tempest* not only reveal one of the roots of our ecological crisis but offer us possible avenues for redressing the balance between humans and nature. These artists disclose the motives and consequences of our "self-proclaimed soliloquist" ways (Manes 22), and provide us with models for reexamining our old myths in order to learn new lessons.

Note

1. A longer revised version of the Césaire portion of this paper appears in *ISLE: Interdisciplinary Studies in Literature and Environment* 3.2 (1996). 47–61.

Works Cited

ARNOLD, JAMES A. *Modernism and Négritude: The Poetry and Poetics of Aimé Césaire.* Cambridge: Harvard UP, 1981.

ASHCROFT, BILL, GARETH GRIFFITHS, AND HELEN TIFFIN. *The Empire Writes Back: Theory and Practice in Post-Colonial Literatures.* London: Routledge, 1989.

BERMAN, MORRIS. *The Reenchantment of the World.* Ithaca: Cornell UP, 1981.

CÉSAIRE, AIMÉ. *A Tempest.* Trans. Richard Miller. Ubu Repertory Theatre Publications, 1985.

———. *Discourse on Colonialism.* Trans. Joan Pinkham. New York: Monthly Review, 1972.

———. "Poetry and Cognition." *Aimé Césaire: L'homme et l'oeuvre.* Kesteloot, Lilyan and Barthélemy Kotchy. Paris: Presses Africaines, 1973.

DE CERTEAU, MICHEL. *The Practice of Everyday Life.* Berkeley: U of California P, 1984.

DEVALL, BILL. "The Deep Ecology Movement." *Natural Resources Journal* 20.1 (1980): 299–322.

ECO, UMBERTO. "A Correspondence on Postmodernism." *Zeitgeist in Babel: The Postmodern Controversy.* Ed. Ingebord Hoesterey. Bloomington: Indiana UP, 1991. 242–53.

GREENAWAY, PETER. *Prospero's Books: A Film of Shakespeare's* The Tempest. New York: Four Walls Eight Windows, 1991.

GREENBLATT, STEPHEN J. "Learning to Curse: Aspects of Linguistic Colonialism in the Sixteenth Century." *First Images of America: The Impact of the New World on the Old.* Ed. Fredi Chiappelli. Vol. 2. Berkeley: U of California P, 1976. 561–80.

MANES, CHRISTOPHER. "Nature and Silence." *The Ecocriticism Reader: Landmarks in Literary Ecology.* Ed. Cheryll Glotfelty and Harold Fromm. Athens: U of Georgia P, 1996. 3–29.

MERLEAU-PONTY, MAURICE. *The Primacy of Perception and Other Essays on Phenomenological Psychology, the Philosophy of Art, History and Politics.* Evanston: Northwestern UP, 1964.

OELSCHLAEGER, MAX. *The Idea of Wilderness: From Prehistory to the Age of Ecology*. New Haven: Yale UP, 1991.

RODGERS, MARLENE. "Prospero's Books. Word and Spectacle: An Interview with Peter Greenaway." *Film Quarterly* 45.2 (1991–92): 11–19.

SHEPARD, PAUL. *Man in the Landscape: A Historic View of the Esthetics of Nature*. 1967. College Station: Texas A&M UP, 1991.

TURMAN, SUZANNA. "Peter Greenaway." *Films in Review* 43 (1992): 105–9.

TYLER, STEPHEN A. "Post Ethnography: From Document of the Occult to Occult Document." *Writing Culture: The Poetics and Politics of Ethnography*. Ed. James Clifford and George E. Marcus. Berkeley: U of California P, 1986. 123–40.

WHITE, LYNN, JR. "The Historical Roots of Our Ecological Crisis." *Science* 155 (1967): 1203–7.

WOLFE, CARY. "Nature as Critical Concept: Kenneth Burke, the Frankfurt School, and 'Metabiology'." *Cultural Critique* 18 (1991): 65–96.

Seeing, Believing, Being, and Acting
Ethics and Self-Representation in Ecocriticism and Nature Writing

H. LEWIS ULMAN

Seeing and Believing

In a posting to the Association for the Study of Literature and Environment e-mail list in December 1994, Daniel Patterson asked participants in the list to consider whether courses on literature and the environment should "be limited to the 'objective' understanding of the cultural forms we analyze" or should "apply a standard of judgment (such as sustainability, wise use, the extension of ethics . . .) in order to determine whether a particular human response to the natural environment . . . is helpful or harmful or beautiful or evil or holy or pernicious or insipid." Of course, Patterson deliberately brackets the terms of his own question, setting aside the ideal of objective understanding, but he leaves in place the tensions among analysis, judgment, and action that are all too familiar to teachers of literature and environment—or, for that matter, to any teacher of literature who combines literary analysis with cultural criticism.

The tension among analysis, judgment, and action in courses on literature and environment is understandable and heuristically valuable. On the one hand, analysis of textual form is a *sine qua non* for literature courses, and class discussions that focus on cultural criticism to the exclusion of textual analysis lose the critical edge provided by close reading. On the other hand, a course in literature and environment that focuses too narrowly and exclusively on the nuances of textual analysis loses much of its reason for being, rooted as such courses generally are in the belief that textual, material, and ideological constructs are bound together in reciprocally enabling and defining relationships. Most of us, I suspect, strive for a constructive engagement of textual analysis and culturally and ecologically situated judgment and action. In a recent course on women's voices in American nature writing, for instance, in addition to asking formal questions about textual representations, my students and I also asked what effects particular ways of reading our texts might have on the oppressive constructions of culture and nature that concern both ecofeminism and deep ecology, were we to embrace the ideology of those readings and act accordingly in the world.

Of course, one can't engage in textual analysis, cultural criticism, or environmental activism for very long in the postmodern academy without confronting the notion that analysis, judgment, situation, and action are inextricably intermingled. But students of literature and environment face an added complication: the aspects of texts that are often of most interest to ecocriticism mirror the critic's own negotiations among these aspects of being-in-the-world. Consider how Carolyn Merchant's caveat about descriptions of the natural world applies equally well to analyses of texts: "It is important to recognize the normative import of descriptive statements about nature. . . . [D]escriptions and norms are not opposed to one another . . . but are contained within each other. Descriptive statements about the world can presuppose the normative; they are then ethic-laden" (4). The relationship is reflexive: normative values and behaviors influence what we see, and our perceptions shape our values and actions. In short, seeing and reporting are always already engaged with believing, being, and acting.

Indeed, some might argue that seeing, believing, being, and acting are virtually coterminous constituents of a socially constructed self and subjectivity. By contrast, I will argue that it is valuable to construct a dynamic, reciprocally defining relationship among these elements. More specifically, I am assuming that the interplay between textual analysis and ideology is in large part what makes critical reading possible and allows us to negotiate meaning and interpretation across differing ideologies and cultural circumstances. Similarly, I assume that the interplay between natural history and environmental ethics enables us to appreciate the complexity of the interrelated physical and cultural systems in which we are all deeply embedded.

To further probe this interplay among seeing, believing, being, and acting in representations of self and nonhuman nature, I turn now to two essays that I teach in my courses on nature writing, Alice Walker's "Am I Blue?" and Richard Nelson's "The Gifts." I am particularly interested in how students respond to the authors' self-representations because I believe that the rhetorical power of much nonfiction nature writing stems from the construction of an ethos that compellingly represents ways of seeing and believing that promise to show us how to be and act at home in the natural world, even when the authors' circumstances, perceptions, and values seem far removed from our own.[1] "Am I Blue?" and "The Gifts" intersect thematically; both explore human attitudes toward nonhuman animals, and both argue eloquently that relationships with other animals should be based on humility and respect. But they diverge in ways that have elicited thoughtful discussion in my class about how we account for, and respond to, subjectivity and self-representation as we construct readings of texts and ethical stances toward nonhuman nature.

"Am I Blue?"

"Am I Blue?" consists of Alice Walker's reflections about her "neighbor"—a white horse named Blue—in a pasture adjoining the house in which she lived during a three-year residence in the country. The narrative thread of this brief essay is spare: Walker and her partner earn Blue's trust by feeding him apples; she learns (or, more precisely, rediscovers) that animals can communicate feelings to us through their gazes; she muses on the similarities between Blue's situation and those of various oppressed human groups; she empathizes with Blue's pain when his partner—a mare named Brown—is removed from the pasture by the horse's owner; and, in a surprising conclusion, she recognizes her own complicity in the underlying dynamics of oppression.

What drives this essay toward its conclusion is less its spare narrative structure than an increasingly complex matrix of ethical agents identified, and in some cases identified with, by the narrator, who implicitly or explicitly conveys approbation or condemnation of the various agents and their actions toward others. In very general terms, the essay examines cruel exercises of power, identifying in many of them a common, underlying pattern in which those in power forget "deep levels of communication" that at one time bound them to the powerless with bonds of respect and love. While she draws analogies to the mistreatment of children by adults, of blacks by whites, of

Indians by settlers, and of women by men, the focus of Walker's essay is the indifference of humans to animals' suffering.

What strikes me as most interesting about the essay, both rhetorically and ethically, is the narrator's self-representation. Through her interactions with Blue in the opening paragraphs of the essay, the narrator appears to have learned a lesson, to have rediscovered "the depth of feeling one could see in horses' eyes" and remembered "that human animals and nonhuman animals can communicate quite well" (4, 5). Though she implicitly contrasts her own kindness to Blue with the thoughtless cruelty of the neighbor's children and the more calculated cruelty of Blue's and Brown's owners, she also implicates her past self in Blue's mistreatment. Describing Blue's grief after Brown is removed summarily from his pasture once they have mated, Walker tells us that she "almost laughed (I felt too sad to cry) to think that there are people who do not know that animals suffer. People like me who have forgotten, and daily forget, all that animals try to tell us. 'Everything you do to us will happen to you; we are your teachers, as you are ours. We are one lesson'" (7). Here we have no self-righteous champion of animal rights, but a black woman who identifies in herself the same blindness to suffering that has contributed to the oppression of women and African Americans in our culture.

It might be easy for the reader to take Walker's inclusion of herself among the human oppressors of animals as a merely rhetorical gesture by someone who, after all, is eloquently defending animals' dignity. By extension, it might be easy for the reader to identify with her enlightened state. But Walker deals the reader a final *coup de main* that turns a spotlight on the insidious workings of forgetfulness and power. Of a discussion with a friend about Blue's predicament, Walker writes with sharp irony: "As we talked of freedom and justice one day for all, we sat down to steaks. I am eating misery, I thought, as I took the first bite. And spit it out" (8). This final scene dramatizes how difficult it can be to transcend our habitual ways of seeing and believing, to enter at least in part into the being of others—particularly nonhuman others—and act accordingly.

"The Gifts"

Richard Nelson's "The Gifts" will at first strike many students as the ethical antipodes of "Am I Blue?" It tells the story of two "moments of grace" experienced during a hunting trip to an island in the Bering Sea off the coast of Alaska, but Nelson represents this hunt as part of a larger effort to shape an environmental ethic based on the traditions of Alaska's Koyukon Indians:

> Traditional Koyukon people follow a code of moral and ethical behavior that keeps a hunter in right relationship to the animals. They teach that all of nature is spiritual and aware, that it must be treated with respect, and that humans should approach the living world with restraint and humility. Now I struggle to learn if these same principles can apply in my own life and culture. Can we borrow from an ancient wisdom to structure a new relationship between ourselves and the environment? Or is Western society irreversibly committed to the illusion that humanity is separate from and dominant over the natural world? (118–19)

The essay's title refers to two encounters with deer, but before presenting those incidents Nelson first establishes his character as a hunter attempting to follow Koyukon tradition and, in some cases, adapt it to his own cultural context. Soon after arriving on the island, he finds a pile of deer bones he had left at the base of a tree the previous winter, "saying they were for the other animals, to make clear that they were not being thoughtlessly wasted" (119). Later, he encounters a doe that he "could almost certainly have taken," but he decides to "wait for a larger buck and let the doe bring on next year's young" (120). When he finally does encounter a large buck, the animal comes so close that Nelson cannot at first bring himself to shoot: "In the Koyukon way, [the deer] has come to me; but in my own he has come too close" (122). However, when the buck sees Nelson and runs, Nelson is caught up in the "predator's impulse to pursue" his prey: "Almost at that instant, still moving without conscious thought, freed of the ambiguities that held me before, *now no less animal than the animal I watch* . . . I carefully align the sights and let go the sudden power. The gift of the deer falls like a feather in the snow" (123; emphasis added). Here Nelson represents himself having transformed his being temporarily in order to kill the deer without the taint of human arrogance. After offering thanks to the deer "for giving itself" to him, Nelson guts the carcass, returns to his cabin, and prepares a meal of venison, wishing his son were there so he could "explain to him again that when we eat the deer its flesh is then our flesh. The deer changes form and becomes us, and we in turn become creatures made of deer"

(125). For all their sacramental presentation, the rituals of this first encounter become but a preamble for the second.

When Nelson sees a doe bedded in the forest the next day, he decides to see how close he can get to her. As he gets to within ten yards of the doe after an excruciatingly slow and careful approach, a buck walks out of the forest and nuzzles the doe. In a dreamlike scene, the doe then turns from the buck and walks up to Nelson, allowing him to touch her on the head before she finally catches his scent and bolts into the woods. Nelson resists trying to explain this transcendent experience: "Was the deer caught by some reckless twinge of curiosity? Had she never encountered a human on this wild island? Did she yield to some odd amorous confusion? I really do not care. I would rather accept this as pure experience and not give in to the notion that everything must be explained" (130). For all his reluctance to explain, however, Nelson is quick to find "vital lessons" in his encounters with the deer that transcend the differences of their outcomes: "These events could be seen as opposites, but they are in fact identical. Both are founded in the same principles, the same relationship, the same reciprocity" (130). The principle, in short, "is to approach all of earth-life, of which we are a part, with humility and respect" (130).

By certain conventions of textual analysis, "Am I Blue?" and "The Gifts" could also be seen as essentially identical narratives, for they espouse similar ethical principles: attend to what animals communicate to us, treat them with humility and respect. But an interpretive turn to such shared ethical principles elides the complexity of these two texts, as does an ideological reading of them as contradictory testimonials for vegetarianism or subsistence hunting. Rather, we need to investigate how two authors who see animals in such similar ways and believe in such similar ethical principles nevertheless come to be and act so differently in the world—in short, how they fashion and represent themselves as ethical agents.

Being and Acting

"Am I Blue?" and "The Gifts," it seems to me, remind us that engaging literature and environmental ethics in the classroom is more complicated than getting students to understand the right facts or adopt the right principles and then expecting them to act accordingly—knowledge and ethical principles are not mapped directly onto our actions. In "Am I Blue?" and "The Gifts," the authors *see* similar features of nonhuman nature: they watch animals carefully and recognize in their expressions and behavior an awareness and communicative capacity most of us overlook. Further, both authors come to *believe* in similar ethical principles of respect for nonhuman animals. But the authors write from very different cultural positions—an African American woman writer and a white, male anthropologist—and they *act* very differently, in one case refusing to eat animal flesh, in the other hunting and eating animals as a transcendent gesture of respect.

Of course one could argue that the circumstances depicted in these essays are entirely different: Walker is reacting against the treatment of livestock and says nothing directly against the sort of hunting Nelson practices; Nelson's meat eating is part of a reflective and ritualized environmental ethic that does not explicitly condone modern meat processing. But noting such differences complicates rather than resolves the ethical issues at stake for readers trying to assess what these texts say *to them* about being and acting at home in the world, for the modes of being and acting represented in the essays may not translate any more easily to readers' lives and circumstances than they do across the essays. By way of example, I can certainly imagine a reader responding to these essays by imagining a more respectful practice of raising and consuming livestock or by more critically examining the cultural and ecological effects of agricultural practices that produce vegetarian foodstuffs.

In other words, to resolve ethical dilemmas *solely* on the basis of abstract ethical principles is to ignore the web of cultural and natural contexts in which ethical decisions are made. As Jim Cheney has argued, "to contextualize ethical deliberation is, in some sense, to provide a narrative, or story, from which the solution to the ethical dilemma emerges as the fitting conclusion" (144). Perhaps this is why writers interested in environmental ethics often turn to nonfiction narrative. For, as Cheney points out, environmental ethics requires situated narratives of self-fashioning, accounts of being and acting at home in the natural world: "the moral point of view wants a storied residence in Montana, Utah, Newfoundland, a life on the tall grass prairie, or on the Cape Cod coastline. . . . Character always takes narrative form; history is required to form character. . . . If a holistic ethic is really to incorporate the whole story, it must systematically embed itself in historical eventfulness" (145). Thus, fictional and nonfictional narratives offer unique contributions to environmental ethics. And if *critical readings* of such narratives are also to contribute to environmental ethics, they must first be embedded in the modes of seeing, believing, being, and acting revealed to the reader through authors' explicit and implicit self-representations, then engaged in dialogue with other narratives of place—including the critic's own. To accomplish all this, ecocriticism needs to pay particular attention to the reciprocally defining relationships among textual analysis, standards of value, and culturally and environmentally situated subjectivity and action.

As a white, heterosexual, male academic, a backsliding vegetarian, and a one-time outdoor educator who has lived and traveled in the far north but settled into a career in academia and life in the city, I have ample opportunities to trace the differences between my own being and actions and those of the personae in "Am I Blue?" and "The Gifts," even as I am drawn into these narratives by similarities to my own past and present experiences and beliefs. For most of my students from rural and urban Ohio, the differences between themselves and the authors' personae are even more apparent, the similarities more difficult to find. But in class, my students and I work against the tendency to lump together modes of seeing, believing, being, and acting—to categorize ethical stances with labels like animal rights advocate, vegetarian, or romantic. Instead, we attempt to read at the borders where the perceptions, values, situations, and actions of the personae in these essays engage one another dynamically. We were chastened by Alice Walker's brutally honest analysis of power and oppression across the boundaries of race, gender, age, and species, but at the same time we recognized that avoiding meat would not necessarily free us from complicity in the misery of animals who are displaced by agriculture or urban development. That project would require analysis of our local circumstances. Similarly, we admired Richard Nelson's willingness to learn from indigenous cultures, but we realized that it would be an unmitigated ecological disaster for millions from the lower forty-eight states to descend upon Alaska—or their own state forests—and take up subsistence hunting. In the final analysis, Nelson's and Walker's self-representations do not fully answer the question of how *we* might borrow from their narratives a way "to structure a new relationship between ourselves and the environment." That is a story we must continually compose, individually and collectively, out of the complex, local practices of seeing, believing, being, and acting that shape our lives and ourselves.

Note

1. For a fuller explication of my argument regarding the role of ethos in nature writing see my essay "'Thinking Like a Mountain': Persona, Ethos, and Judgment in American Nature Writing."

Works Cited

CHENEY, JIM. "Eco-Feminism and Deep Ecology." *Environmental Ethics* 9 (1987): 115–45.

MERCHANT, CAROLYN. *The Death of Nature: Women, Ecology, and the Scientific Revolution*. New York: Harper & Row, 1983.

NELSON, RICHARD K. "The Gifts." *On Nature: Nature, Landscape, and Natural History*. Ed. Daniel Halpern. San Francisco: North Point P, 1987. 117–31.

PATTERSON, DANIEL. Posting to the Association for the Study of Literature and Environment e-mail list. December 2, 1994.

ULMAN, H. LEWIS. "'Thinking Like a Mountain': Persona, Ethos, and Judgment in American Nature Writing." *Green Culture: Environmental Rhetoric in Contemporary America*. Ed. Carl G. Herndl and Stuart C. Brown. Madison: U of Wisconsin P, 1996. 46–81.

WALKER, ALICE. "Am I Blue?" *Living by the Word: Selected Writings, 1973–1987*. New York: Harcourt Brace, 1988. 3–8.

Don DeLillo's Postmodern Pastoral
DANA PHILLIPS

A decade after its publication, the contribution of Don DeLillo's *White Noise* to our understanding of postmodern cultural conditions has been thoroughly examined by literary critics (see, for example, the two volumes of essays on DeLillo's work edited by Frank Lentricchia). The novel has been mined for statements like "Talk is radio," "Everything's a car," "Everything was on TV last night," and "We are here to simulate"—statements that critics, attuned to our culture's dependence on artifice and its habit of commodifying "everything," immediately recognize as postmodern slogans. What has been less often noticed, and less thoroughly commented on, is DeLillo's portrait of the way in which postmodernity also entails the devastation of the natural world.

Frank Lentricchia, in his introduction to the *New Essays on White Noise*, has pointed out that "The central event of the novel is an ecological disaster. Thus: an ecological novel at the dawn of ecological consciousness" (7). But Lentricchia does not develop his insight about the "ecological" character of the novel. Neither does another reader, Michael Moses, who in his essay on *White Noise*, "Lust Removed from Nature," argues that "postmodernism, particularly when it understands itself as the antithesis rather than the culmination of the modern scientific project, confidently and unequivocally banishes from critical discussion the questions of human nature and of nature in general" (82). Moses does not pursue this point, but I would argue that one of the great virtues of DeLillo's novel is the thoroughgoing and imaginative way in which *White Noise* puts the questions not just of human nature but of "nature in general" back on the agenda for "critical discussion."

The dearth of commentary on DeLillo's interest in the fate of nature is explained, not just by the fact that contemporary literary critics tend to be more interested in the fate of culture, but also by the fact that one has to adjust one's sense of nature radically in order to understand how, in *White Noise*, natural conditions are depicted as coextensive with, rather than opposed to, the malaise of postmodern culture. This adjustment is not just a task for the reader or critic: it is something the characters in the novel have to do every day of their lives.

As a corrective to the prevailing critical views of the novel, *White Noise* might be seen as an example of what I will call the *postmodern pastoral*, in order to foreground the novel's surprising interest in the natural world and in a mostly forgotten and, indeed, largely bygone rural American landscape. At first glance the setting of the novel and its prevailing tone seem wholly unpastoral. But then the pastoral is perhaps the most plastic of modes, as William Empson demonstrated in *Some Versions of Pastoral*. The formula for "the pastoral process" proposed by Empson—"putting the complex into the simple" (23)—is one which might appeal to the main character and narrator of *White Noise*, Jack Gladney. Gladney is someone who would like very much to put the complex into the simple, but who can discover nothing simple in the postmodern world he inhabits, a world in which the familiar oppositions on which the pastoral depends appear to have broken down. And thus the postmodern pastoral must be understood as a *blocked* pastoral—as the expression of a perpetually frustrated pastoral impulse or desire. In qualifying my assertion that *White Noise* is an example of postmodern pastoral in this way, I am trying to heed Paul Alpers's warning that "modern studies tend to use 'pastoral' with ungoverned inclusiveness" (ix). However, Alpers's insistence that "we will have a far truer idea of pastoral if we take its representative anecdote to be herdsmen and their lives, rather than landscape or idealized nature" (22) would prevent altogether the heuristic use of the term I wish to make here. With all due respect to herdsmen, the interest of the pastoral for me lies more in the philosophical debate it engenders about the proper relation of nature and culture and less in its report on the workaday details of animal husbandry or the love lives of shepherds.

Jack Gladney is not a shepherd, but a professor of Hitler Studies at the College-on-the-Hill, which is situated in the midst of an unremarkable sprawl of development that could be called "suburban," except that there is no urban center to which the little town of Blacksmith is subjoined. Like almost everything else in *White Noise*, the town, to judge from Jack Gladney's description of it, seems displaced, or more precisely, unplaced. Jack tells us that "Blacksmith is nowhere near a large city. We don't feel threatened and aggrieved in quite the same way other towns do. We're not smack in the path of history and its contaminations" (85). He proves to be only half-right: the town is, in fact, subject to "contaminations," historically and otherwise. Jack's geography is dated: Blacksmith is not so much "nowhere" as it is Everywhere, smack in the middle—if that is the right phrase—of a typically uncentered contemporary American landscape of freeways, airports, office parks, and abandoned industrial sites. According to Jack, "the main route out of town" passes through "a sordid gantlet of used cars, fast food, discount drugs and quad cinemas" (119). We've all run such a gantlet; we've all been to Blacksmith. It is the sort of town you can feel homesick for "even when you are there" (257).

Thus, despite a welter of detail, the crowded landscape in and around Blacksmith does not quite constitute a *place*, not in the sense of "place" as something that the characters in a more traditional novel might inhabit, identify with, and be identified by. Consider Jack's description of how Denise, one of the Gladney children, updates her "address" book: "She was transcribing names and phone numbers from an old book to a new one. There were no addresses. Her friends had phone numbers only, a race of people with a seven-bit analog consciousness" (41). Consciousness of place as something that might be geographically or topographically (that is, locally) determined has been eroded by a variety of more universal cultural forms in addition to the telephone. Chief among them is television—Jack calls the TV set the "focal point" of life in Blacksmith (85). These more universal cultural forms are not just forms of media and media technology, however; the category includes such things as, for example, tract housing developments.

Despite the prefabricated setting of *White Noise* and the "seven-bit analog consciousness" of its characters, an earlier, more natural and more pastoral landscape figures throughout the novel as an absent presence of which the characters are still dimly aware.

Fragments of this landscape are often evoked as negative tokens of a loss the characters feel but cannot quite articulate, or more interestingly—and perhaps more postmodern as well—as negative tokens of a loss the characters articulate, but cannot quite feel. In an early scene, one of many in which Jack Gladney and his colleague Murray Jay Siskind ponder the "abandoned meanings" of the postmodern world (184), the two men visit "THE MOST PHOTOGRAPHED BARN IN AMERICA," which lies "twenty-two miles into the country around Farmington" (12). In his role as narrator, Jack Gladney often notes details of topography with what seems to be a specious precision. But the speciousness of such details is exactly the issue. Even though it is surrounded by a countrified landscape of "meadows and apple orchards" where fences trail through "rolling fields" (12), Farmington is not at all what its name still declares it to be: a farming town. The aptness of that placename, and of the bits of rural landscape still surrounding the barn, has faded like an old photograph. As Murray Jay Siskind observes, "Once you've seen the signs about the barn, it becomes impossible to see the barn" (12). The reality of the pastoral landscape has been sapped, not just by its repeated representation on postcards and in snapshots, but also by its new status as a tourist attraction: by the redesignation of its cow paths as people-movers. The question of authenticity, of originality, of what the barn was like "before it was photographed" and overrun by tourists, however alluring it may seem, remains oddly irrelevant (13). This is the case, as Murray observes, because he and Jack cannot get "outside the aura" of the cultural fuss surrounding the object itself, "the incessant clicking of shutter release buttons, the rustling crank of levers that advanced the film" (13)—noises that drown out the incessant clicking of insect wings and the rustling of leaves that once would have been the aural backdrop to the view of the barn.

As the novel's foremost authority on the postmodern, Murray is "immensely pleased" by THE MOST PHOTOGRAPHED BARN IN AMERICA (13). He is a visiting professor in "the popular culture department, known officially as American environments" (9), an official title that signals the expansion of the department's academic territory beyond what was formerly considered "cultural." Jack dismisses Murray's academic specialty as "an Aristotelianism of bubble gum wrappers and detergent jingles" (9)—that is, as a mistaken attempt to uncover the natural history of the artificial. Jack finds the BARN vaguely disturbing.

But *White Noise* is about Jack's belated education in the new protocols of the postmodern world in which he has to make his home. Jack learns a lot about those protocols from Murray and his colleagues, one of whom lectures a lunchtime crowd on the quotidian pleasures of the road (arguably a quintessentially postmodern American "place"). Professor Lasher sounds something like Charles Kuralt, only with more attitude:

> "These are the things they don't teach," Lasher said. "Bowls with no seats. Pissing in sinks. The culture of public toilets. The whole ethos of the road. I've pissed in sinks all through the American West. I've slipped across the border to piss in sinks in Manitoba and Alberta. This is what it's all about. The great western skies. The Best Western motels. The diners and drive-ins. The poetry of the road, the plains, the desert. The filthy stinking toilets. I pissed in a sink in Utah when it was twenty-two below. That's the coldest I've ever pissed in a sink in." (68)

Lasher's little diatribe may seem to suggest that DeLillo is satirizing the much-heralded replacement of an older cultural canon by a newer one: Lasher would throw out the Great Books, if he could, in favor of "the poetry of the road." But in *White Noise* it is not so much the replacement as it is the displacement of older forms by newer ones, and the potential overlapping or even the merger of all those forms in an increasingly crowded cultural and natural landscape, that DeLillo records. "The great western skies," the "Best Western motels," "the road, the plains, the desert"—all are features of a single, seamless landscape.

Because of their ability to recognize so readily the odd continuities and everyday ironies of the postmodern world, the contentious members of the department of American environments seem better-adapted than their more cloistered colleagues. Their weirdness is enabling. By pursuing their interest in and enthusiasm for things like the culture of public toilets, they collapse the distinction between the vernacular and the academic and shorten the distance between the supermarket, where tabloids are sold, and the ivory tower, where the library is housed. It is instructive that whenever one of their more extreme claims is challenged, members of the department tend to reply in one of two ways: either they say, "It's obvious" (a refrain that runs throughout the novel), when of course it (whatever it may be) isn't at all obvious. Or they simply shrug and say, "I'm from New

York." In *White Noise*, all knowledge is local knowledge, but one must understand how shaped by the global the local has become. We're all from New York.

While it is true that we can "take in"—as the saying goes—a landscape, the literal ingestion of nature (that is, of discrete bits and selected pieces of it) is probably the most intimate and most immediate of our relations with it. In a telling passage from the opening pages of the novel, Jack and his wife Babette encounter Murray Jay Siskind in the generic food products aisle of the local supermarket:

> His basket held generic food and drink, nonbrand items in plain white packages with simple labeling. There was a white can labeled CANNED PEACHES. There was a white package of bacon without a plastic window for viewing a representative slice. A jar of roasted nuts had a white wrapper bearing the words IRREGULAR PEANUTS. (18)

What is striking about the contents of Murray's cart is the way in which, despite the determined efforts of all those labels to say in chorus the generic word FOOD, they seem to be saying something else entirely. These "nonbrand items" actually seem to be all brand, nothing but brand; their categorical labels seem like mere gestures toward the idea of food, evocations of its half-forgotten genres. Remember UNCANNED PEACHES? Visible bacon? REGULAR PEANUTS? The packaging and the labels do not resolve the question of contents. They raise it; that is, they heighten it, so that it seems more important than ever before.

The jar of IRREGULAR PEANUTS in particular has a disturbing, perhaps even slightly malign quality, as Murray explains: "'I've bought these peanuts before. They're round, cubical, pock-marked, seamed. Broken peanuts. A lot of dust at the bottom of the jar. But they taste good. Most of all I like the packages themselves. [. . .] This is the last avant-garde. Bold new forms. The power to shock'" (19). Siskind's identification of the jar of peanuts as part of "the last avant-garde" suggests that cultural production has reached the *ne plus ultra* of innovation, that henceforward it will consist not in making things new, but in the repackaging of old things, of the detritus of nature and the rubble of culture. "Most of all," Murray says, "I like the packages themselves." So there will not be any more avant-gardes after this one—it is not the latest, but "the last." Those IRREGULAR PEANUTS mark the end of history: more than just irregular, they are

APOCALYPTIC peanuts. No wonder Murray savors them. Each is a bite-size reminder of the "end of nature" and the "end of history," two of the postmodernist's favorite themes.

The CANNED PEACHES, the invisible bacon, and the IRREGULAR PEANUTS also demonstrate very clearly how postmodern culture does not oppose itself to nature (as we tend to assume culture must always do). Instead, it tries to subsume it, right along with its own cultural past. But one would like to protest that despite all this repackaging and attempted subsumption, the fact is that peanuts—even IRREGULAR ones—do not result from cultural production, but from the reproduction of other peanuts. One wants to say that natural selection (plus a little breeding), and not culture, has played the central and determining role in the evolution of peanuts of whatever kind. But the role of nature as reproductive source, even as an awareness of it is echoed in certain moments of the novel, tends to get lost in the haze of cultural signals or "white noise" that Jack Gladney struggles and largely fails to decipher, probably because *all* noise is white noise in a postmodern world. Murray Jay Siskind, as a connoisseur of the postmodern, is sublimely indifferent to factual distinctions between, say, the natural and the cultural of the sort that still worry less-attuned characters like Jack Gladney.

That they must eat strange or irregular foods is only part of the corporeal and psychological adjustment Jack and his family find themselves struggling to make. At least they remain relatively *aware* of what they eat, in that they choose to eat it. But "consumption" is not necessarily always a matter of choice in *White Noise*: there are things that enter the orifices, or that pass through the porous membranes of the body, and make no impression on the senses. These more sinister invaders of the body include the chemicals generated by industry, many of them merely as by-products, chemicals that may or may not be of grave concern to "consumers"—not entirely the right term, of course, since few people willingly "consume" toxins. After all, we do not have to eat the world in order to have intimate relations with it, since we take it in with every breath and every dilation of our pores. This suggests that the much-bewailed runaway consumerism of postmodern society is not the whole story: there are other kinds of exchange taking place that do not necessarily have to do with economics alone. The cash nexus is certainly economic, but the chemical nexus is both economic and ecological; the economy of by-products, of toxic waste, is also an ecology. Economic or ecological fundamentalism makes it hard to tell the whole story about postmodernism, as DeLillo is trying to do.

During the novel's central episode, the "airborne toxic event," Jack Gladney is exposed to a toxin called Nyodene Derivative ("derivative" because it is a useless by-product). Nyodene D and its possible effects are first described for Jack by a technician at the SIMUVAC ("SIMUVAC" is an acronym for "simulated evacuation") refugee center: "'It's the two and a half minutes standing right in it that makes me wince. Actual skin and orifice contact. This is Nyodene D. A whole new generation of toxic waste. What we call state of the art. One part per million can send a rat into a permanent state'" (138–39). The technician's last phrase is richly ambiguous: does "a permanent state" mean death or never-ending seizure or a sort of chemically induced immortality? This ambiguity terrifies Jack, and he begins to seek some surer knowledge of the danger he is in. At this point in the narrative, DeLillo's novel speaks most clearly about the effect the postmodern condition has on our knowledge of our bodies (and thus on our knowledge of nature). Having crunched all Jack's numbers in the SIMUVAC computer, the technician informs him, "I'm getting bracketed numbers with pulsing stars," and he adds that Jack would "rather not know" what that means (140). Of course, that is precisely what Jack would most like to know. The attempt at clarification offered by the technician at the end of their conversation does nothing to explain to Jack exactly when, why, and how he might die: "It just means that you are the sum total of your data. No man escapes that" (141).

The remainder of the novel is taken up with Jack Gladney's attempt to escape the reductive judgment of his fate given by the SIMUVAC technician and his computer (whose bracketed numbers with pulsing stars "represent" Jack's death, but do so opaquely, in a completely nonrepresentative way, rather like the white package marked BACON that conceals the supermarket's generic pork product). As the repository of junk food and as a host for wayward toxins and lurking diseases, Jack's body has become a medium, in much the same way that television or radio are media. His postmodern body is hard to get at in the same way that the nameless voices on television—the ones that throughout the novel say macabre things like "Now we will put the little feelers on the butterfly" (96)—cannot always be identified, much less questioned or otherwise engaged in dialogue. In *White Noise*, the body itself is mediated, occult, hard to identify, and unavailable for direct interrogation by any solely human agent or agency. The postmodern body is, then, a curiously disembodied thing. It no longer makes itself known by means of apparent symptoms that can be diagnosed by a doctor, nor by means of feelings that can be decoded by the organism it hosts (it may be a little old-fashioned to think of this organism as a "person"). During his interview with Dr. Chakravarty, Jack utters a tortured circumlocution in response to the simple question, "How do you feel?". His carefully qualified reply, "To the best of my knowledge, I feel very well," demonstrates how distant from him Jack's body now seems (261). That this body just happens to be his own gives Jack no real epistemological advantage. In a postmodern world, technology and the body are merely different moments of the same feedback loop, just as the city and the country are merged in a common landscape of death. Because it is the place in which distinctions between bodies and machines, and between the city and country, have collapsed, "Autumn Harvest Farms" is an exemplar of postmodern pastoral space: at Autumn Harvest Farms, the machine not only *belongs* in the garden, it *is* the garden.

However confused he may be, and however paralyzed by his half-living, half-dead condition, Jack Gladney does seem to "feel," at times, a certain lingering nostalgia about and interest in "nature in general." This longing, if not for the prelapsarian world, then at least for some contact with a nature other than that of his own befuddled self, is apparent even in the lie Jack tells the Autumn Harvest Farms clinician in response to a question about his use of nicotine and caffeine: "Can't understand what people see in all this artificial stimulation. I get high just walking in the woods" (279). The only time in the novel when Jack actually goes for something like a walk "in the woods" is when he visits a rural cemetary. Like everything else in the novel, this cemetary has an overdetermined quality: it is called "THE OLD BURYING GROUND," and it is both authentic—actually an old burying ground, that is—and a tourist trap. It is both what it is and an image or metaphor of what it is. And so THE OLD BURYING GROUND seems uncanny, with the same kind of heightened unreality about it that gives Murray's jar of IRREGULAR PEANUTS and THE MOST PHOTOGRAPHED BARN IN AMERICA their peculiar auras.

Nonetheless, it may be at the old burying ground that Jack comes closest to feeling some of the peace that the countryside can bring:

> I was beyond the traffic noise, the intermittent stir of factories across the river. So at least in this they'd been correct, placing

> the graveyard here, a silence that had stood its ground. The air
> had a bite. I breathed deeply, remained in one spot, waiting to
> feel the peace that is supposed to descend upon the dead, wait-
> ing to see the light that hangs above the fields of the landscapist's
> lament. (97)

But in this remnant of an older, more pastoral landscape set in the midst of a contemporary sprawl—across the Lethean river separating the graveyard from the factories in town, but still sandwiched between the town, the freeway, and the local airport—Jack does not quite have the epiphany he is so clearly seeking. His hope of living within the natural cycle of life and death suggested to him by his visit to THE OLD BURYING GROUND has already been foreclosed by events. Direct encounter with nature, "walking in the woods," is no longer possible, not only because nature seems to have become largely an anecdotal matter of broadcast tidbits of information about animals (bighorn sheep, dolphins, etc.), but also because nature, like the body, has been ineluctably altered by technology. THE OLD BURYING GROUND, landscaped as it is, and given its purpose, is a crude example of this alteration, however comforting Jack finds it.

The supermarket is the place that the characters in the novel depend on most for a sense of order, pattern, and meaning, and thus it fulfills something of the cultural function that used to be assigned to the pastoral. The difference is that the supermarket has an obscure relationship to the rest of the world, particularly to the natural world whose products it presumably displays. The supermarket is a pastoral space removed from nature. Unfortunately, even this artificial haven is disturbingly altered by the novel's end: "The supermarket shelves have been rearranged. It happened one day without warning. There is agitation and panic in the aisles, dismay in the faces of the older shoppers" (326). The "agitation and panic in the aisles" of the super-market links the postmodern condition back to an older set of fears and confusions that predate the repose that the pastoral is supposed to offer. DeLillo makes this very clear earlier in the novel when he has Jack Gladney use the word "panic" to describe his anxiety upon awakening in the middle of the night: "In the dark the mind runs on like a devouring machine, the only thing awake in the universe. I tried to make out the walls, the dresser in the corner. It was the old defenseless feeling. Small, weak, deathbound, alone. Panic, the god of woods and wilderness, half goat" (224). Thus Jack finds himself in the wilderness even while he is supposedly safe at home in Blacksmith. The order and rationality, the civilized space, that modernity (like the pastoral) supposedly created seems to be no longer a feature of the postmodern landscape.

The postmodern pastoral, unlike its predecessors, cannot restore the harmony and balance of culture with nature, because the cultural distinctions that the pastoral used to make—like that between the city and the country—have become too fluid to have any force and are dissolved in the toxic fog of airborne events. Neither culture nor nature are what they used to be. But perhaps DeLillo's point is that they never were, that the distinction between culture and nature cannot be taken as an absolute. As a novelist, he knows just how thoroughly "all of culture and all of nature get churned up again every day" (2), as Bruno Latour puts it in his appositely-titled book, *We Have Never Been Modern* (from which it follows that we cannot possibly be "postmodern" in the strict sense of the term). DeLillo is also aware of another point on which Latour insists: he realizes that the everyday churning up of nature and culture is not just a matter of media representations. Latour argues that "the intellectual culture in which we live does not know how to categorize" the "strange situations" produced by the interactions of nature and culture because they are simultaneously material, social, and linguistic, and our theories are poorly adapted to them (3). They are not cognizant of what Latour likes to call "nature-culture."

It seems to me that Latour—and DeLillo—are right, and that postmodernist theorists (unlike postmodern novelists, whose work is often finer grained than theory) have invested too much in the ultimately false distinction between nature and culture. They have tried to argue what amounts to a revision of Frederick Jackson Turner's frontier thesis, first promulgated in his 1893 essay, "The Significance of the Frontier in American History." Turner argued that the closing of the frontier and the disappearance of wilderness was a turning point in American culture; the postmodernists—especially the more radical or pessimistic postmodernists like François Lyotard and Jean Baudrillard, or Fredric Jameson—argue that the disappearance of nature is a turning point in global culture. Postmodernism is a frontier thesis for the next millenium, more dependent on what has been called "the idea of wilderness" than its exponents have realized.

Works Cited

ALPERS, PAUL. *What Is Pastoral?* Chicago: U of Chicago P, 1996.
DELILLO, DON. *White Noise*. New York: Viking Penguin, 1985.
EMPSON, WILLIAM. *Some Versions of Pastoral*. Norfolk: New Directions Books. n.d.
LATOUR, BRUNO. *We Have Never Been Modern*. Trans. Catherine Porter. Cambridge: Harvard UP, 1993.
LENTRICCHIA, FRANK, ED. *Introducing Don DeLillo*. Durham: Duke UP, 1991.
———. *New Essays on White Noise*. Cambridge: Cambridge UP, 1991.
MOSES, MICHAEL. "Lust Removed from Nature." In *Introducing Don DeLillo*. Ed. Frank Lentricchia. Durham: Duke UP, 1991: 63–86.
TURNER, FREDERICK JACKSON. "The Significance of the Frontier in American History." In *The Frontier in American History*. Tucson: U of Arizona P, 1986. 1–38.

"The World Was the Beginning of the World"
Agency and Homology in A. R. Ammons's Garbage

LEONARD M. SCIGAJ

Contemporary environmental poetry can enliven our perceptions and instruct us about how to live in harmony with the energies and the ecological cycles of the planet that produced us. But, as Glen Love has recognized, contemporary critics have most often ignored environmental literature (201–5). Environmental poetry can gain a more serious hearing from contemporary literary critics, and thus gain a wider audience, by engaging in the poststructural debate over language and agency. Paul Smith, in *Discerning the Subject*, delineated just how limited the various poststructural theoretical projects are when it comes to providing an empowering inwardness that can lead to emancipatory social change. Smith revealed how, under the theoretical constructs of Derrida, Lacan, and Barthes, the subject is largely a passive construct interpolated or conscripted into predetermined social roles. Once the subject as a child absorbs the language of his or her culture, with all of its attendant assumptions, that subject's decision making is forever mediated and determined by the social roles and behavioral codes inscribed in that language by the prevailing culture. If this is the case, how can the subject become the resisting agent capable of developing a unique subjectivity that can lead to self-empowerment and social change? Charles Altieri summarized the central dilemma surrounding poststructural understandings of language and agency:

> [T]he entire [poststructural] enterprise is haunted by the models of subjection developed by Lacan and Althusser—the one in terms of family romance, the other in terms of the interpollation that gives agents a place in the social order. Both perspectives prove disturbingly essentialist, asserting necessary structures for the psyche which divide subjective agency from itself and force it to live its imaginative life in thrall to some other that it internalizes as the means of attributing to itself powers of subjectivity. . . . Derrida and Foucault are more abstract, but perhaps also even more trapped, since for them the depersonalizing force is a property of all categories: categories and hence concepts force third-person frameworks necessary for intelligibility on first-person states and thus necessarily banish subjective agency to the margins of a public world. (221–22)

Altieri tried to establish a measure of subjective agency for postmodern poetry by applying Wittgensteinian categories—deictic shifters such as "now" and "this"—which, in the work of postmodern poets John Ashbery and C. K. Williams, appear to create a self-reflexive consciousness of how one wills the self and disposes the self toward what is being told, rather than by idealizing what actually gets told. The postmodern establishment poet Jorie Graham has developed her own poetics, in some ways similar to Altieri's emphasis on self-reflexivity, for creating a liberating freedom from historical determinism. She glorifies desire at the moment of composition. Graham equates closure with the imperialist, manifest-destiny desire to own reliable utilitarian truths; hence in her poems she endeavors to delay closure by finding linguistic and formalist ways of keeping open the gap between signified concepts and referential truth (Graham and Gardner 79–104).

Yet for environmental poets the problem with such enterprises as Altieri's and Graham's is that, in developing various amalgams of Derrida, Barthes, Lacan, and Wittgenstein for their theoretical superstructure, they valorize language as their ultimate model and guide. Their a priori assumption is that humans know the world only through language and, no matter how unreliable, "the limits of *language*," according to Wittgenstein's *Tractatus*, "mean the limits of my world" (5.62). Postmodern establishment poets such as John Ashbery, Jorie Graham, and Robert Hass assume a dualism where the space of the poem contains only the subjective musings of the anthropocentric psyche engaged in addressing problems relevant only to humans.

They place the planet on the table constructed by the human imagination, as does Wallace Stevens in his late poems, whereas environmental poets want to restore a healthy sense of humans as a subset of nature, dependent for their survival on the health of natural ecosystems. Environmental poets want to downsize that table and direct our attention toward the larger and more powerful planet that grounds and sustains it.

Hence environmental poets such as A. R. Ammons, Wendell Berry, W. S. Merwin, Gary Snyder (and a host of others including Brendan Galvin, Joy Harjo, Denise Levertov, Mary Oliver, Robert Pack, Marge Piercy, Adrienne Rich, and Pattiann Rogers) find language to be simply a tool or exchange mechanism—a secondhand product that since its creation has been utilized by dominant societies for the subjection of subordinate societies, in every case producing ecological degradation. The cycles of nature, not the much more limited laws of language, are the models we must conform to, or risk elimination. Language is useful only to the extent that it can orient us away from anthropocentric domination by helping us recognize a biocentric view in which humans are but one of many species that must cooperate to create a sustainable future.

A. R. Ammons's National-Book-Award-winning ecopoem *Garbage* (1993) is a prime example of an environmentalist work informed by the ecological cycles of nature, not the laws of language. The book-length poem is ostensibly about the huge I–95 landfill outside Miami, which Ammons presents as a ziggurat, a religious edifice of American culture, attended by high priests driving bulldozers, spreaders, and garbage trucks (8, 98). Yet the text is as serious as it is comic; in it one finds a critique of anthropocentric uses of language and a series of homologies constructed to orient our perceptions toward greater harmony with the cycles of nature. As with *Tape for the Turn of the Year* (1965), Ammons wrote *Garbage* on an adding machine tape, though one slightly wider than that used for the earlier poem (63). The tape's material form emphasizes its purpose—to suggest continuities among the elements and beings of the natural world. Near the center of the text, Ammons bluntly articulates a critique of any poetics that uses language as its model. The referential world is the origin and support for all language:

> . . . the world was the beginning
> of the world; words are a way of fending in the
> world: whole languages, like species, can

> disappear without dropping a gram of earth's
> weight, and symbolic systems to a fare you well
> can be added without filling a ditch or thimble (50–51)

Later in this passage Ammons asserts his affinities with animals, in this case birds, because he "shares with them, their states // of being and feeling" (51).

Throughout its lengthy tape, *Garbage* emphasizes homologous shared relationships between humans and various orders of sentient life. Rather than valorizing human subjectivity by emphasizing its self-referential desires along a linguistic chain of signifiers, the work stresses subjective agency through recognizing in a living language our shared biocentric relationships with all orders of sentient beings who live in harmony with our planetary ecosystems. In contrast, a "Styrofoam" language emphasizes an anthropocentric will to power (74):

> I know the entire language of chickens,
> from rooster crows to biddy cheeps: it is a
>
> language sufficient to the forms and procedures
> nature assigned to chicken-birds but a language,
>
> as competition goes, not sufficient to protect
> them from us: our systems now
>
> change their genes, their forms and procedures,
> house them up in all-life houses, trick their
>
> egg laying with artificial days and nights:
> our language is something to write home about:
>
> but it is not the world: grooming does for
> baboons most of what words do for us. (51–52)

The satiric barb at the end of this passage emphasizes that the cart of language theory, with postmodern theorists in the driver's seat, must not be placed before the natural horse, its referential source and enabling power.

In *Postmodernism, or the Cultural Logic of Late Capitalism* (1991), Fredric Jameson asserted that one can recover from the structuralists the Lévi-Straussian use of homology without subscribing to any transcendent concept of "structure" (187–93). Homologies for Jameson bring in the rich substance of the natural world; they "imply analogies between objects, content, or raw materials within discourse" and thus question the theories of postmodern language critics who assume the ontological primacy of language (239). Indeed, Lévi-Strauss in *Totemism* argued that the tribal mind in early human culture used metaphor and analogy to grasp shared differences among species, and thus mediated the limits of the natural world and the needs of human culture without the dualistic split between humans and nature that Christianity created (3, 78–82, 91). For Lévi-Strauss the homological method of finding structural relationships that mediate discontinuities is both a "universal feature of human thinking" and *the* essential method of structural anthropology; its purpose is to apprehend the union, the integration, of method and reality, where the opposites are not Derridean linguistic binaries, but those of human and nonhuman thinking and social organization (90–91). Lévi-Strauss even suggested that homologous thinking, like much tribal philosophy, presents the intuition, like Bergson's *élan vitale*, that all entities on the planet are homologously interrelated because they are "materialized forms of creative energy" (98).

The word "homology" derives from the Greek *homologous* and the Latin *homologia*. It means "agreeing." Today, in the physical sciences such as biology and chemistry, homologies denote shared characteristics or corresponding structural features among families of compounds, organs, or creatures deriving from a common ancestor. Unlike analogies, where two essentially dissimilar entities share *one* common characteristic, homologies suggest *many* corresponding structural features shared by two entities. I would like to suggest that Ammons uses a homologous method in *Garbage* to close the postmodern gap between signification and referentiality, and to reveal the wondrous and beautiful interrelations of the human and nonhuman worlds that share nature's creative energy.

Those familiar with Ammons's earlier work recognize that much of it is built around an ecological application of Plotinus's One and Many dialectic, as Ammons himself once asserted (Ammons and Walsh 111). In *Garbage* Ammons enriches his grasp of the pre-Socratics by noting that their theories are based upon homologous relationships: "things are sustained by interrelations and // variety" (40). As he announces that *Garbage* is "a scientific poem, // asserting that nature models values" (20), he freely plays with the contiguities of *Garbage* "disposal" and the poetic "disposition" or formal arrangement of the poem's flow. He reveals that "the poem" is "about the pre-socratic idea of the // dispositional axis from stone to wind, wind / to stone" (20). Nature, not language, is the referent and model for the poem's structural logic, and Ammons's pre-Socratic predecessor now appears to be Anaximenes, who argued that the underlying substance of reality was something one, infinite, and determinate: air. Of a practical and scientific bent, Anaximenes believed that nature's forms are interrelated, composed of degrees of rarified or condensed air. This is a marvelously biocentric view: all earthly entities are made of one common "stuff" in rarefied or condensed forms. In its most rarefied form air is fire, and air slowly descends into more condensed and less active forms of energy from winds to stones (Warner 16–18; Nahm 65–67).

For Ammons nature offers us active agency, a "globe-round selfempowerment" whose basis is a continual flow of transforming energy (*Garbage* 49). In a perfectly homologous way, the poet recycles language by incinerating dead language in the heat of compositional activity: "in the poet's mind dead language is hauled / off to and burned down on, the energy held and // shaped into new turns and clusters, the mind / strengthened by what it strengthens" (20). Again nature, not language, is the model for a series of homologous interrelationships:

> nature, not
> we, gave rise to us: we are not, though, though
> natural, divorced from higher, finer configurations:
>
> tissues and holograms of energy circulate in
> us and seek and find representations of themselves
>
> outside us, so that we can participate in
> celebrations high and know reaches of feeling
>
> and sight and thought that penetrate (really
> penetrate) far, far beyond these our wet cells (21)

Ammons's style in *Garbage*, homologous with nature and his method of formal composition, "wraps back round" from the high to the low, the sublime to the scatological, giving us not only heaven but "heaven's daunting asshole" (22). He strives for a "chunky intermediacy" between language that apprehends the spiritual and a "gooey language" like the "bird do" that once dabbed his shoulder. Always his poetic energy bubbles with typical Ammonsesque buoyancy: "undone by / do, I forged on," and while "forging" or composing he notes homologous relationships between the "bellyroundworms" in the "bird do" and the life he creates in the moment of composition.

The bathetic "bird do" meditation then wraps around to the spiritual, creating in the space of a few unrhymed couplets a Taoist circuit of emanations from a shared source:

> if you've derived from life
> a going thing called life, life has a right to
>
> derive life from you: ticks, parasites, lice,
> fleas, mites, flukes, crabs, mosquitoes, black
>
> flies, bacteria: in reality, reality is like
> still water, invisible, spiritual: the real
>
> abides, spiritual, while entities come and go (98–99)

Incorrigibly comic, Ammons less than a page later meditates on why men often measure themselves by the size of their penises and women by the size of their breasts. But we seldom find ideal, perfect representatives of male or female beauty in nature, argues Ammons, because nature "likes a broad spectrum approaching disorder so // as to maintain the potential of change with / variety and environment." The ideal resides "implicit and stable" amid the homologous variations, the "shorties" and "flopsies" (100–101).

As Ammons expatiates airily on subjects that range from the trivial to the profound, he rarefies and condenses his meditations in ways that mimic Anaximenes's understanding of nature's movements. Thus Ammons can "mix [his] motions in with the mix of motions" (84). Like the proverbial blabbermouth, Ammons talks through and around so many topics that inevitably discussion wraps around to its opposite. In the process the blabbermouth poet contextualizes his meditations, creating a "common place" enriched with homologous references, just as the gyrations of earthworms enrich the soil by tilling it (78–79). Any homely experience becomes a homily on nature's interrelations. In section eleven, for instance, Ammons relates a visit to the Lake Cayuga farmers' market. As he observes the aging participants—the "wobble-legged," the "toothless, big-bellied, bald, broad-rumped"—he realizes that visiting a farmers' market is a ritual sharing and bonding of humanity that renews life. It is humanity "at our best, not killing, scheming, abusing," but renewing life, just as the "huge beech by the water" exfoliates yearly. Then the poet rarefies his discussion to reach for a moment of faith in the human ability to share, to achieve the togetherness of human community. This human ability to share and bond he calls "the magical exception // to the naturalistic rule" (71). Here Ammons immediately perceives a homology that leads him to recognize that the bonding of humans at such pedestrian events is also a survival tactic, a bit of "science // knowledge, craft" quite within nature's laws. Such human gatherings are homologous to the webworm's ability to camouflage itself and blend its body with the coloration of honeysuckle bushes. As the shared knowledge of planting and harvesting at a farmers' market reinvigorates our ties with nature and merges the isolated individual with the group, so nature gives the webworm a purplish streak down its back "exactly the color" of the buds of "honeysucklebushlimb-stems," and colors his feet "exactly the color of the lateralhoneysuck-lebush / limbstems" to allow him to merge with his environment. Humans share their skills at adapting to the environment at a farmers' market just as the webworm, dangling in air on its thread of web, parades its adaptive skills (71–72).

At times Ammons uses his homologous method to refute anthropocentric heresies. In one short passage in which he deliberates upon how the chemistry of potato starch enlivens his chemistry, Ammons refutes three kinds of logical errors: that of dualists who, like postmodern establishment critics, assume an initial divorce of humans from nature; that of developers of aesthetic systems based on the polysemous instability of language; and, that of inflators of human consciousness who assume that nature does not exist without a human consciousness to perceive it:

> the world is not a show consciousness can pull
>
> off or wipe out: because consciousness can neither
> wipe out nor actualize it it is not a show but
>
> the world: if one does not eat perception-blasted
> potato, one will blast perception by the loss of
>
> perception: starch (in Arch) in the potato
> meets with my chemistry to enliven my chemistry,
>
> clear my eyes, harden, perhaps, my muscle, wag
> my tongue (almost certainly): hallelujah: if
>
> death is so persuasive, can't life be: it is
> fashionable now to mean nothing, not to exist,
>
> because meaning doesn't hold, and we do not exist
> forever; this *is* forever, we are now in it (87–88)

Ammons is most sardonic when he uses homology to underscore how far humans have departed from sustainability. For Ammons the culprit is our inflated, anthropocentric trust in language. In section twelve he meditates on the annual accumulation of waste in the carbon cycle. Though animals such as chipmunks live off the plethora of "sugarmaple" seeds that fall each spring, most seeds will turn to detritus without producing trees. But compacted detritus ultimately turns into life-nourishing soil. This process has continued for millennia. Yet since our hunter-gatherer ancestors evolved language about six thousand years ago, "things have gone poorly for the / planet" (74). We have too often used our words in a utilitarian endeavor that disrupts life-sustaining planetary cycles. Since Sumerian cuneiform and Alexandrian vellum, humans have created "surface-mining words" that, though "intricate as the realities they represent," have gradually polluted and destroyed ecosystems:

>   all this *Garbage!* all
>
> these words: we may replace our mountains with
> trash: leachments may be our creeks flowing
>
> from the distilling bottoms of corruption:
> our skies, already browned, may be our brown
>
> skies: fields may rise from cultivation into
> suffocation
> . . . .
>   we have replaced
> the meadows with oilslick: when words have
>
> driven the sludge in billows higher than our
> heads—oh, well, by then words will have left
>
> the poor place behind: we'll be settling
> elsewhere or floating interminably, the universe
>
> a deep place to spoil (75–76)

Ammons milks his sardonic mood to its bitter end at the conclusion of section twelve, where he suggests that ultimately humans will leave this planet taking only "the equations, cool, lofty, eternal, that were // nowhere here to be found when we came" (76). Ammons mocks our anthropocentric self-inflation by carrying the logic to its ad nauseam conclusion in the final couplet of the section, illustrating the apocalyptic end of "styrofoam verbiage" (74): "we'll kick the *l* out of the wor*l*d and cuddle / up with the avenues and byways of the word" (77).

Postmodern establishment poets derive their poetics from the language theory of Derrida, Lacan, Barthes, and Wittgenstein—from those who privilege textuality and believe that all experience is mediated by language. These poets desire to write in the gaps between concepts and the referential world, to write with what Heidegger would call an originary language in the flash moment of creation, without the stale meanings of sedimented language (57–108). But this gap too easily becomes a cushy limbo, a safe anthropocentric refuge from the pressures and difficulties of living in the world that nature created. This position is a newer New Criticism, another attempt to valorize poetic experience as ontologically distinct from ordinary language and experience, as Lentricchia has shown in his discussion of the critical line running from Coleridgean genius through New Critical formalism and the aesthetics of Murray Krieger (213–54). This is the position where, according to Lentricchia, "we are trapped in the prison-house of language; we have access to no substance beyond language but only to words in their differential and intrasystematic relations to one another. We have no right, after such assumptions, to say anything except that all utterances are fictions which never break out of their fictionality" (220).

In section nine of *Garbage*, Ammons devises a parable of the rabbit and the chipmunk that may highlight the anthropocentric self-indulgence of modeling one's poetics on language. To stay alive, the "sniffy rabbit" must recognize both his nakedness in the world and his intimate connectedness to his environment. Hence he often moves just far enough away to survive the pounce of a natural predator such as the local tabby cat. In this middle distance the rabbit stands stone still, blending into his surroundings so intimately and yet so alertly that the predators don't notice him. The chipmunk, on the contrary, is all bursts of undirected energy followed by periods of lolling about on the concrete in sunlight. So into himself as to be nearly disengaged from his immediate environment, he is unaware of the stalking tabby, who soon "in / thrusting gulps and crunches down[s] chippy" (59). Ecodisasters may gulp down humans at an ever faster rate the longer we continue to loll myopically in anthropocentric preoccupations with language. The earth is crammed full, avers Ammons; tread it warily and gain "selfempowerment" by cooperating with nature.

Ammons's *Garbage* offers answers to two very serious scholarly questions. Lentricchia asked "is it not urgent to grasp what one believes to be the actual interconnections in human experience and refuse to detain oneself by meditating upon admittedly fictional entities?" (221). In his "Introduction" to *The Environmental Imagination*, Lawrence Buell asked an equally anguished question: "Must literature always lead us away from the physical world, never back to it?" (11). Ammons's homologous method in every section of *Garbage* leads us back toward the referential world and its interrelations, to a world where (pace Blake) even the *Garbage* is holy. With a sly grin Ammons intimates that, unless we redirect our attention away from linguistic systems and attend to the health of our ecosystems, we may not have a pot or a planet for our retirement petunias.

Works Cited

ALTIERI, CHARLES. "Contemporary Poetry as Philosophy: Subjective Agency in John Ashbery and C. K. Williams." *Contemporary Literature* 33 (1992): 214–41.

AMMONS, A. R. *Garbage*. New York: Norton, 1993.

———. *Tape for the Turn of the Year*. Ithaca: Cornell UP, 1965.

———, AND WILLIAM WALSH. "An Interview with A. R. Ammons." *Michigan Quarterly Review* 28 (1989): 105–17.

BUELL, LAWRENCE. *The Environmental Imagination: Thoreau, Nature Writing, and the Formation of American Culture*. Cambridge: Harvard UP, 1995.

GRAHAM, JORIE, AND THOMAS GARDNER. "An Interview with Jorie Graham." *Denver Quarterly* 26 (Spring 1992): 79–104.

HEIDEGGER, MARTIN. *On the Way to Language*. Trans. Peter D. Hertz. New York: Harper & Row, 1982.

JAMESON, FREDRIC. *Postmodernism, or the Cultural Logic of Late Capitalism*. Durham: Duke UP, 1991.

LENTRICCHIA, FRANK. *After the New Criticism*. Chicago: U of Chicago P, 1980.

LÉVI-STRAUSS, CLAUDE. *Totemism*. Trans. Rodney Needham. Boston: Beacon P, 1963.

LOVE, GLEN A. "Revaluing Nature: Toward an Ecological Criticism." *Western American Literature* 25 (1990): 201–15.

NAHM, MILTON C., ED. *Selections from Early Greek Philosophy*. 3d ed. New York: Appleton-Century-Crofts, 1947.

SMITH, PAUL. *Discerning the Subject*. Minneapolis: U of Minnesota P, 1988.

WARNER, REX. *The Greek Philosophers*. New York: New American Library, 1958.

WITTGENSTEIN, LUDWIG. *Tractatus Logico-Philosophicus*. Trans. D. F. Pears and B. F. McGuinness. London: Routledge & Keegan Paul, 1961.

Notes on Contributors and Editors

CHRIS BEYERS is an instructor at Auburn University. He has published a number of articles examining the contexts of American literature and the aesthetics of twentieth-century poetry. His current research investigates how poetry of colonial Maryland reacts to social, political, and economic forces.

MICHAEL P. BRANCH is an Associate Professor of Literature and Environment at the University of Nevada, Reno, where he is cofounder and past president of the Association for the Study of Literature and Environment. A specialist in early American environmental literature, he is coeditor of *The Height of Our Mountains: Nature Writing from Virginia's Blue Ridge Mountains and Shenandoah Valley* (Johns Hopkins University Press, 1998).

ANNA CAREW-MILLER is the head of the English Department at the Gunnery, an independent secondary school in Connecticut, where she teaches nature writing courses and outdoor education. She continues to write and publish articles on Mary Austin in particular, and on women's nature writing in general.

JONI ADAMSON CLARKE is an Assistant Professor of English at the University of Arizona, Sierra Vista. Her book on Native American literature and ecological literary theory is forthcoming from the University of Arizona Press.

J. GERARD DOLLAR is a Professor of English at Siena College and has published articles on Cather, Hardy, Conrad, and Stevenson. He has recently received a Fulbright lectureship to teach literature and the enviroment at the University of Turku in Finland.

KELLY M. FLYNN is a Ph.D. candidate in English at Princeton University. The project, "Peaceable Kingdoms: Constructions of

Animal Life in Nineteenth-Century Literature," addresses an epistemological quandry which arose in the post-Darwinian era, when the fabulist tradition came into conflict with the possibility of animal autonomy; and issues of language, being, religion, and ecology turned around the re-appropriation of animal metaphor. Authors treated include Hawthorne, Thoreau, Wilder, Frost, and White.

WILLIAM HOWARTH is a Professor of English at Princeton University, where he teaches courses in American literature and environmental history. He is the author of twelve books and more than a hundred chapters and articles on American literature, including *The Book of Concord: Thoreau's Life as a Writer* (Viking, 1982). He serves on the editorial boards of *Environmental History* and of *ISLE*, and is an associate of the Natural Rural Studies Council.

ROCHELLE JOHNSON is completing her dissertation at the Claremont Graduate University. Her research focuses on traditions of nature writing in nineteenth-century America. She guest edited a special issue of *Women's Studies* in 1996 titled "Women and Nature." She is coeditor of Susan Fenimore Cooper's *Rural Hours* (University of Georgia Press, 1998).

ANN E. LUNDBERG is a Ph.D. candidate in English at the University of Notre Dame, where she is completing a dissertation titled "mapping the Geologic Wilderness: Science, Nature Writing, and the American Self." She has served as Managing Editor of the journal *Nineteenth Century Contexts* and currently teaches the study and practice of nature writing at Notre Dame.

DAVID MAZEL is an Assistant Professor of English at Adams State College in Alamosa, Colorado. His most recent book is *Mountaineering Women: Stories by Early Climbers* (Texas A&M University Press, 1994). He is currently completing a critical study of colonial and nineteenth-century American environmental writing and editing an anthology of early ecocriticism.

MICHAEL MCDOWELL is a faculty member in the English Department of Portland Community College in Oregon. He teaches American Literature, composition, and creative writing.

ANNE E. MCILHANEY, an Associate Professor of English at Webster University, recently completed her doctorate in English at the University of Virginia, where she wrote her dissertation on English Renaissance fishing literature.

ROBERT MELLIN is a Ph.D. candidate and lecturer in the English Department at Wayne State University. His dissertation project, "Outside Theory: Speculations of the Grounds of Literary Ecology and US Culture," is being funded by Wayne State University Humanities Center Fellowship. He is coediting a forthcoming essay collection on Gayatri Chakravorty Spivak and cultural studies.

JOHN P. O'GRADY is the author of *Pilgrims to the Wild* (University of Utah Press, 1993) and *Grave Goods* (forthcoming from University of Utah Press). He teaches literature at Boise State University.

DANIEL PATTERSON is an Assistant Professor of English at California State University, San Bernadino. He is a Colonialist, specializing in the literature and culture of seventeenth-century New England. Recently his attention has turned to the history of writing about nature in North America. He is coeditor of Susan Fenimore Cooper's *Rural Hours* (University of Georgia Press, 1998).

DANIEL J. PHILIPPON is Assistant Professor of rhetoric at the University of Minnesota, Twin Cities. He is coeditor of *The Height of our Mountains: Nature Writing from Virginia's Blue Ridge Mountains and Shenandoah Valley* (Johns Hopkins University Press, 1998) and public relations coordinator of the Association for the Study of Literature and Environment.

DANA PHILLIPS has taught American literature at the University of Pennsylvania, is Princeton University, and Bryn Mawr College. He is the author of articles on postmodernism and the environment.

STEPHANIE SARVER whose Ph.D. is from the University of California, Davis, is a writer living in San Mateo, California. Her scholarly interests include relationships between business and environment, particularly the way capitalism shapes scientific research. Her book *Uneven Land: Nature and Agriculture in American Writing* is forthcoming from the University of Nebraska Press.

LEONARD M. SCIGAJ is an Associate Professor of English at Virginia Polytechnic Institute and State University. He is an expert on twentieth-century British and American environmental poetry and has completed a manuscript entitled *Sustainable Poetry: Four American Ecopoets.*

DEBORAH SLICER is Associate Professor of Philosophy at the University of Montana. Her work focuses on environmental philosophy, and she also co-directs a student-run organic garden that donates its harvest to low-income Missoulians.

SCOTT SLOVIC is an Associate Professor of Literature and Environment and director of the Center for Environmental Arts and Humanities at the University of Nevada, Reno. The founding president of the Association for the Study of Literature and Environment, and the current editor of the journal *ISLE: Interdisciplinary Studies in Literature and Environment,* he has coedited *Being in the World: An Environmental Reader for Writers* (Macmillan, 1993) and *Literature and the Environment: Writings on Humans in Nature* (Addison Wesley Longman, 1998), and is the author of *Seeking Awareness in American Nature Writing* (University of Utah Press, 1992).

ERIC TODD SMITH is a Ph.D. candidate in English at the University of California, Davis, where he teaches literature and composition. He is currently writing the Western Writer's Series pamphlet on Gary Snyder's *Mountains and Rivers Without End,* to be published in the spring of 1999.

JOHN TALLMADGE is a Professor of Literature and Environmental Studies at the Union Institute Graduate School in Cincinnati, Ohio. A scholar of American nature writing and the recent president of the Association for the Study of Literature and Environment, he is the author of *Meeting the Tree of Life: A Teacher's Path* (University of Utah Press, 1997).

H. LEWIS ULMAN is an Associate Professor of English at Ohio State University, where he teaches courses on writing, literature, rhetoric, and literacy. His publications include books and articles on eighteenth-century British philosophy and rhetoric as well as essays on American nature writing and computers and literacy. His current research focuses on the study of natural history in eighteenth-century Britain and on the rhetoric of American nature writing.

PAULA WILLOQUET-MARICONDI is a Ph.D. candidate in Comparative Literature and Film Studies at Indiana University. Her dissertation offers an ecocritical reading of the work of filmmaker and artist, Peter Greenway. Her research explores film and environment, deep ecology, and postmodernism; her articles have appeared in such journals as *ISLE, Studies in Twentieth-Century Literature, Film/Literature Quarterly, Cinema Journal,* and *Literature/Interpretation/Theory.*

Index

Abbey, Edward, 103–4; misogyny in the works of, 97–99, 104
Abram, David, 38 (n. 2)
Alexander, Meena, 76
Alpers, Paul, 236
Altieri, Charles, 247–48
"Am I Blue?" *See* Walker, Alice
American Cancer Society, 109, 116 (n. 1)
American Indian literature. *See* Native American literature
Ammons, A. R.: *Garbage* as response to postmodern view of language and nature, 249–57
Ammons, Elizabeth, 91, 93 (n. 16)
Anaximenes, 252, 253
Anderson, Chris, 20
anglers, women: images of in literature, 55–63
Ashbery, John, 248
Association for the Study of Literature and Environment, The (ASLE), xiv, 51, 225
Audubon, John James, xvi, 15; autobiographical aspects of his nature writing, 119–27
Austin, Mary, 45–46, 48; how her representations of nature reveal her sense of place, 79, 82–92; public and private voices of 80–82, 83, 85–86, 91; mother's influence on, 79–81, 82, 86, 91

Baillie, Joanna, 68, 69
Bakhtin, Mikhail, 25–26, 33
Bardach, Ann, 9
Barthes, Roland, 141
Bartram, John, 15
Bate, Jonathan, 67
bears: grizzly, 114–15, 116 (n. 3)
Berg, Peter, 27 (n. 2)
Berman, Morris, 218, 220
Berners, Juliana, 57, 63
Berry, Wendell, 113, 249
biophilia, xiii–xiv. *See also* desire
Blithedale Romance, The. See Hawthorne
Bodega Head, 41–44, 50–51
body, the human: as bioregion, 112–14. *See also* women
Bordo, Susan, 111, 114
Breen, Jennifer, 68, 69
Bromell, Nicholas, 145
Browne, William, 64 (n. 8)
Buell, Lawrence, xi–xii, 257
Burke, Kenneth, 216
Burroughs, John, 93 (n. 9), 189

Callicott, J. Baird, 14
Campbell, SueEllen, xiv, 15

Cather, Willa, 6, 98; *My Antonia*, 4–6, 7; misogyny in the works of, 99–101, 104

Catlin, George, xvi; his environmentalism implicated in nationalist and imperialist politics, 129–31, 134, 140; and ecological identity, 137–41

Césaire, Aimé: *A Tempest* as animistic re-reading of Shakespeare's *The Tempest*, 209, 210–14, 222; *Notebook of a Return to the Native Land*, 211; on the role of the poetry, 213, 217. *See also* Greenaway; Shakespeare

Cheney, Jim, 231

Chielens, Edward E., 191 (n. 10)

Cicardo, Barbara J., 122

Cockburn, Alexander, 14

Cole, Thomas, 149

composition. *See* ecocomposition

Cummings, Glenn N., 191 (n. 8)

Curran, Stuart, 68, 69

Curtis, George William, 147–48

Davis, Robert Con, 72, 73

de Certeau, Michel, 216, 217

Delafosse, Maurice, 211

DeLillo, Don: *White Noise* as postmodern pastoral, 235–45

Dennys, John, 57, 58, 59

desire, xiii–xiv; in Mary Austin's work, 88–91. *See also* biophilia

Devall, Bill, 212

Dillard, Annie, xvi, 8; *Pilgrim at Tinker Creek* and the excursion form, 198–202; *Pilgrim* contrasted with *Refuge* (Williams), 203–7

Dixon, Terrell F., 20

Donne, John, 59, 60, 61, 62

Eco, Umberto, 214

ecocomposition, 19–26; definition of, 20, 22; texts, 20–21, 22, 24

ecocriticism: origins and purposes of, xi–xiv, 29–30, 31; in academia, 3, 7–8; guidelines for courses in, 7–8, 225–26, 231–33; and issues of race and human rights, 10, 14–16; and the subject-status of nature, 30–38; romantic ecology, 67–68. *See also* "transformative ecocriticism"

ecofeminism, 12, 16, 107. *See also* women

ecological identity, 134–40

"ecologicality." *See* ecological identity

Eliade, Mircea, 48, 49

Ellis, Amanda, 68

Emerson, Ralph Waldo, xvi, 43, 153; his philosophy of how the farmer relates to nature, 155–62

Empson, William, 236

environmental poetry, 247–49

Fiennes, Celia, 55–56, 62, 63

Ford, Alice, 119, 127

Forkner, Ben, 122

Fromm, Harold, xi

Garbage. *See* Ammons, A. R.

Garber, Frederick, 175 (n. 1)

"Gifts," "The." *See* Nelson

Gilmore, Michael T., 147, 163 (n. 4)

Ginsberg, Allen, 49

Glotfelty, Cheryll, xi, 29–30, 31, 32, 34, 36–37

Godwin, William, 74

Graham, Jorie, 248–49

Grant, Anne, 69

Graulich, Melody, 82

Greenaway, Peter, 209; *Prospero's Books* as ecological re-reading of Shakespeare's *The Tempest*, 209, 210, 214–21, 222. *See also* Césaire; Shakespeare

Hallowell, Christopher, 20

Haraway, Donna, 32, 36, 37

Hass, Robert, 248–49

Hawthorne, Nathaniel, xvi; on the relation between manual and mental labor, 146–48; *The Blithedale Romance* as expression of his relation to the physical world, 150–53

Hecht, Susanna, 14

Heidegger, Martin, 256

Hemingway, Ernest, 98

Heraclitus, 43

Hoby, Margaret, 55, 63
Hooker, Thomas, 46
Huck Finn, 98–99
Hunter, Susannah. *See* Austin: mother's influence on

Innocent Epicure, 61

Jameson, Fredric, 250–51
Jaycox, Faith, 87
Jefferson, Thomas, 158
Jenseth, Richard, 20
Johnson, Linck C., 187

Kaufman, Wallace, 24
Kazin, Alfred, 179
Kolodny, Annette, xv, 104–5 (n. 3), 122–23
Kraitsir, Charles, 167
Kroeber, Karl, 70

landscape painting, 148–49
Latour, Bruno, 35–36, 37, 245
Lentricchia, Frank, 235, 256–57
Lévi-Strauss, Claude, 251
Levy, Walter, 20
literature and environment. *See* ecocriticism
Lopez, Barry, xiii, 15–16, 207
Lotto, Edward E., 20
Love, Glen, 31, 33, 247
Lovelace, Richard, 60
Luther, Martin, 44
Lyon, Thomas J., 197

Maclean, Norman, 8, 98; misogyny in the works of, 101–3, 104
Maine Woods, The. *See* Thoreau
Manes, Christopher, 38 (n. 2)
Marcos, Subcommandante, 9–10
Markham, Gervase, 58
Marlowe, Christopher, 63 (n. 8)
Martin, Emily, 111
Mayan Indians, 9–10
McGann, Jerome J., 76 (n. 1)
McKibben, Bill, 135, 137–41
meditation, 46
Merchant, Carolyn, xv, 226
Merwin, W. S., 249

Miller, James E., Jr., 186
misogyny: in versions of the American West, 97–104. *See also* Abbey; Cather; Maclean
Morgan, Sarah, 20
Moses, 51
Moses, Michael, 235
Muir, John, 48
Murphy, Patrick, 10–11, 32–34, 36

Nabholtz, John, 71
Nash, Roderick, 38 (n. 1), 132, 133, 134
Native American literature, 11, 14, 16
nature writing: as a genre, xiii, 10–11, 12, 15
Nelson, Richard K., xvii, 227, 228–32
New, Elisa, 123
Nuttall, Thomas, 15

Oelschlaeger, Max, 210, 218
O'Grady, John, 89
Okerstrom, Dennis, 20
Ong, Walter, 176 (n. 8)
Oregonian, The, 21
Orion magazine, 24
Orr, David W., xii
Ortiz, Simon, 10, 15–16; "That's the Place the Indians Talk About," 11–12, 14

Pacific Gas and Electric, 42
pastoral: elegy, 4; British piscatory, 56–57, 58–63; Wordsworthian, 67, 75; postmodern, 235–45
Patterson, Daniel, 225
Pearce, Roy Harvey, 131–32, 134
Pilgrim at Tinker Creek. *See* Dillard
piscatory tradition. *See* pastoral: British piscatory
Plant, Judith, 107
Plumwood, Val, 107
postmodernism, 247–49. *See also* pastoral: postmodern

Quigley, Peter, 34–35

Reddy Kilowatt, 44, 50–51
Refuge: An Unnatural History of Family and Place. *See* Williams, Terry Tempest

Rich, Adrienne, 108
Ripley, George, 145, 163 (n. 5)
Robinson, Mary, 69
Roe, Nicholas, 70
romantic ecology. *See* ecocriticism
Ross, Andrew, 135, 136, 141–42 (n. 1)
Ross, Carolyn, 20
Round, Phillip, 176 (n. 7)
Ruether, Rosemary Radford, 16
Runciman, Lex, 20

Sanders, Scott Russell, 119–20, 126–27
Seager, Joni, 135, 141–42 (n. 1)
Shakespeare, William: *The Tempest*, 209, 219–20. *See also* Césaire; Greenaway
Shasta Nation Bioregional Gathering, 46–48
Shepard, Paul, 212
Silko, Leslie Marmon, 10, 15–16; *Almanac of the Dead*, 12–14, 15
Slovic, Scott H., 20, 136
Smith, Charlotte, 68
Smith, Paul, 247
Snyder, Gary, 45, 249
Snyder, William, 71
Specimen Days. See Whitman
Stanley, Thomas, 59, 60
Stevens, Wallace, 46, 249
Stovall, Floyd, 186

Tallmadge, John, 175 (n. 5)
Tempest, The. See Shakespeare
Thomashow, Mitchell, 135, 136–37
Thoreau, Henry David, xvi, 6–7, 151, 197; *The Maine Woods* as exploration of relation between word and world, 165–66, 168–75; *Walden*, 165; linguistic density of, 166–68
"transformative ecocriticism," 10, 16
Turner, Frederick Jackson, 245
Tyler, Stephen, 215, 222

Valenti, Peter, 20

Wagner, Richard, 101
Walden. See Thoreau
Walker, Alice, 227–28, 230–32
Walker, Cheryl, 92 (n. 2)
Walker, Melissa, 20, 22
Waller, Edmund, 56, 57, 59, 60–61, 62, 63
Walton, Izaak, 56, 57–58, 59
Warren, Karen, 12, 16
Warren, Robert Penn, 122
Welty, Eudora, 122
West, Michael, 167
Whalen, Philip, 43, 45
White, Gilbert, 197
White, Lynn, Jr., 210, 212
White Noise. See Delillo
Whitman, Walt, xvi, 44; *Specimen Days* as consideration of relation between humans and nature, 179–89
Whitney, John, 62–63
Williams, C. K., 248
Williams, Patricia, 110, 112
Williams, Terry Tempest, xvii; *Refuge* and the excursion form, 202–5, 207; *Refuge* contrasted with *Pilgrim* (Dillard), 203–7
Wilson, E. O., xiii
"wise use" movement, 24
Wittgenstein, Ludwig, 248
Wolfson, Susan, 70
Wollstonecraft, Mary, 74–75
women: objectification of, 107–12. *See also* body
Woolley, Hannah, 56, 63
Wordsworth, Dorothy, 67–76; influence on brother and other environmental writers, 67–68, 73; and the picturesque, 70–73, 75; nonanthropocentric bioregionalism in the journals of, 72–74; and woman's domestic place, 74–75
Wordsworth, William: unacknowledged indebtedness to sister and other women writers, 67–70; appropriation of sister's voice, 75–76
Wyatt, David, 87

Zapatista National Liberation Army, 9–10, 14, 16